The Physical Chemistry of Membranes

The Physical Chemistry of Membranes

Michael E. Starzak
Department of Chemistry
State University of New York at Binghamton
Binghamton, New York

1984

ACADEMIC PRESS, INC.

(Harcourt Brace Jovanovich, Publishers)

Orlando   San Diego   San Francisco   New York   London
Toronto   Montreal   Sydney   Tokyo   São Paulo

7227-069X

CHEMISTRY

ACADEMIC PRESS, INC.
Orlando, Florida 32887

United Kingdom Edition published by
ACADEMIC PRESS, INC. (LONDON) LTD.
24/28 Oval Road, London NW1 7DX

Library of Congress Cataloging in Publication Data

Starzak, Michael E.
    The physical chemistry of membranes.

    Includes index.
    1. Membranes (Biology) 2. Chemistry, Physical
organic. 3. Biological chemistry. I. Title.
QH601.S68   1983   574.87'5   83-4628
ISBN 0-12-664580-9

PRINTED IN THE UNITED STATES OF AMERICA

84 85 86 87   9 8 7 6 5 4 3 2 1

Contents

Preface   ix

1. Early Physical Techniques

   1. Early History   1
   2. Membrane Molecules   2
   3. Some Surface Properties of Lipids   6
   4. The Gorter–Grendel Experiment   10
   5. Surface Tension   11
   6. Surface Tension Measurements on Membrane Systems   12
   7. X-Ray Scattering by Membranes   15
   8. Birefringence Measurements   16
     Problems   19

2. Thermodynamics

   1. The First Law of Thermodynamics   21
   2. Entropy   24
   3. Legendre Transforms   27
   4. Maxwell's Relations   30
   5. The Gibbs Adsorption Isotherm   31
   6. Boltzmann Statistics   33
   7. The Langevin Equation   36
   8. The Grand Canonical Partition Function and Membrane
     Binding   39
     Problems   43

3. Electrical Interactions

1. Electrical Units   44
2. Alternative Forms of Coulomb's Law   45
3. Electric Dipoles   48
4. Dipole Rotation in Electric Fields   52
5. Dipole–Dipole Interactions   53
6. The Effects of Thermal Agitation   57
7. Induced Electrostatic Interactions   58
8. Capacitance   60
9. Image Charges   62
   Problems   64

4. Solution and Surface Thermodynamics

1. Solution Chemical Potentials   66
2. The Osmotic Pressure   70
3. The Donnan Equilibrium   75
4. Donnan Equilibria for Polyvalent Ions   78
5. The Electrochemical Potential   79
6. Electrical Potentials across Membranes   81
7. The Nerve Action Potential   83
8. The Relative Roles of Ions   84
9. The pH Electrode . 86
10. Ultracentrifugation   88
    Problems   90

5. Ion Transport

1. The Electrode Interface   91
2. Electrodes   94
3. Interfacial Potentials   97
4. Ion Motions in Solution   100
5. Concentration Cells   103
6. The Diffusion Potential   107
7. The Henderson Equation   109
   Problems   113

6. Electronics

1. Input and Output Impedance   114
2. Operational Amplifiers   116
3. Operational Amplifier Summer   118
4. Noninverting Amplifier   121
5. Operational Amplifier Integrators   123
6. Voltage Clamp   128

7. Logic Chips   132
8. Other Integrated Circuit Modules   135
9. Signal Averaging   136
10. Lock-In Amplification   139

7. Membrane Capacitance

1. Resistance–Capacitance Networks   141
2. Sinusoidal Inputs for Equivalent Circuits   146
3. The Measurement of Electrical Parameters   150
4. Cole–Cole Plots   153
5. Complex Dielectric Constants   156
6. The Debye Relaxation Time   161
7. The Capacitance for Suspensions of Cells   164
8. Membrane Capacitance for Conducting Cells   170
   Problems   172

8. The Electrical Double Layer

1. The Helmholtz Double Layer   174
2. The Gouy–Chapman Theory   178
3. The Grahame Equation   184
4. Modification of the Double-Layer Models   188
5. Screening   192
6. Electrocapillarity   195
7. Electrophoresis and Electroosmosis   198
8. Streaming Potential   201
   Problems   204

9. Diffusion

1. The Nature of Diffusion   206
2. Force-Flux Relations   208
3. Fick's First Law of Diffusion   209
4. Fick's Second Law and the Equation of Continuity   211
5. Diffusion across a Membrane   213
6. The Nernst–Planck Equation   218
7. The Goldman Constant Field Equation   220
8. The Goldman–Hodgkin–Katz Equation   224
9. Flux Ratios   227
10. Discrete State Diffusion Models   230
11. The Goldman and Parlin–Eyring Models   234
    Problems   236

10. Irreversible Thermodynamics

1. Reversible and Irreversible Thermodynamics   238
2. Generalized Forces and Fluxes   240
3. The Dissipation Function   243
4. Chemical Reactions   246
5. Electrokinetic Effects   250
6. Osmotic Flow   253
7. Saxen's Relationships   257
8. The Curie Symmetry Principle   259
9. Steady State Coupling   262
   Problems   265

11. Kinetics

1. Basic Kinetics   267
2. Relaxation Kinetics   272
3. Kinetic Activation Energy   273
4. Transition-State Theory   275
5. Enzyme Kinetics   278
6. Graph Theory Kinetics   281
7. Cyclic Kinetic Schemes   284
8. Passive Membrane Transport   287
9. Ion Transport through Channels   292
10. Ion Transport in Multisite Channels   294
11. Non–Steady-State Graph Kinetics   298
    Problems   301

12. Membrane Excitability

1. The Action Potential Revisited   302
2. The Voltage Clamp   302
3. The Hodgkin–Huxley Equations—Preliminary Considerations   307
4. The Hodgkin–Huxley K^+ Equation   310
5. The Hodgkin–Huxley Na^+ Equation   314
6. Effects of Polyvalent Ions   317
7. Gating Currents   319
8. The Cole–Moore Shift   320
9. Ascending Potential Ramps   324
   Problems   325

References   327

Index   329

Preface

This book had its nucleus in a one-semester course on the physical chemistry of membranes. The course is taken by juniors, seniors, and first-year graduate students from our chemistry, biology, and biochemistry programs. Student interest in this interdisciplinary area has been increasing steadily. At the same time, our standard physical chemistry courses assumed an increased emphasis on quantum mechanics and spectroscopy while sacrificing some topics in the physical chemistry of solutions and related topics. In such circumstances, this book was designed to cover topics involving the physical chemistry of membranes.

All of the topics covered in the book are ultimately related to membranes. I have tried to avoid a structure that presents individual topics without emphasizing the relationships among the different topics. The membranes provide the common thread that permits the development of such relationships. For example, the electrical potential that can be generated across a membrane in the presence of a concentration gradient is explored in Chapter 4. In Chapter 9, I develop flux equations for transmembrane ion flow. By considering the zero flux limit, the equilibrium electrical potential is determined for comparison with the equilibrium thermodynamic result of Chapter 4.

Flux relationships are developed in a similar way. Chapter 2 defines the necessary therodynamic variables and the methods of manipulating them. In Chapter 9, these intensive and extensive thermodynamic variables are converted into the forces and fluxes necessary to develop the diffusion equations. In Chapter 10 on irreversible thermodynamics, we develop a more general description of forces and fluxes. Chapter 11 then describes the calculation of such fluxes in more complex membrane systems.

One major advantage of a study of membrane systems lies in the reduction of dimensionality of the system. For processes that proceed perpendicular to the surface, for example, transmembrane diffusion, we often assume homogeneity in directions parallel to the surface. For this reason, three-dimensional problems reduce to one-dimensional problems. The resultant differential equations are then more tractable. They permit the student to make the connections between the physical phenomena and their related mathematical descriptions. This facilitates understanding of the more complex multidimensional systems.

The choice of one-dimensional problems accentuates the overall approach of the book. It is not intended as a monograph to cover all topics in the field of membrane biophysics; the field is expanding too rapidly to effectively do that. Rather, the book is intended as a bridge between topics in an elementary physical chemistry course and continuing developments in the field. I have selected topics that will fill this gap most effectively while providing the necessary mathematical tools for the student to pursue more sophisticated problems.

I must express my thanks to those students who have taken this course with me over the years. Their questions and insights induced me to try new approaches and explanations of the material. I have been able to document those approaches that ultimately proved most lucid in this book.

I also wish to thank my wife, Anndrea, and my children, Jocelyn and Alissa, for providing their time and encouragement with me so I could finish this book.

The Physical Chemistry of Membranes

Chapter 1

Early Physical Techniques

1. EARLY HISTORY

From the time cells were first observed under a microscope, they were observed to have very definite boundaries. The cell retained a definite shape and, although animal cells were generally more flexible than plant cells, it seemed reasonably obvious that a barrier of some sort surrounded the cell and prevented the loss of the material within the cell to the surrounding medium. Some simple experiments also tended to substantiate this observation. If the concentration of the external solution were changed, the cell could be induced to increase or decrease in volume. Such observations were consistent with the notion of a water flow across the boundary.

Even such observations, however, could not establish whether the membrane existed as a distinct entity. Alternatively, the protoplasmic material within the cell might simply coalesce into a structural boundary where it came into contact with the aqueous medium. It became necessary to establish a distinction between these two possibilities.

The boundary region between the cell and the aqueous medium was known to be composed of organic materials. Overton (1899) had established a direct correlation between the amount of organic material that diffused into cells from the external medium and the partition coefficient of the material between an aqueous solution and an oil phase (Fig. 1.1), The more readily the molecules dissolved in oil, the more readily they dissolved or penetrated into the cell. However, the cells were also quite permeable to water, although this solvent could not dissolve in the relevant organic medium, suggesting an added complexity for the membrane system.

Plowe (1930) punctured cells and observed the holes formed in this manner. She concluded that these holes enlarged elastically. They opened too

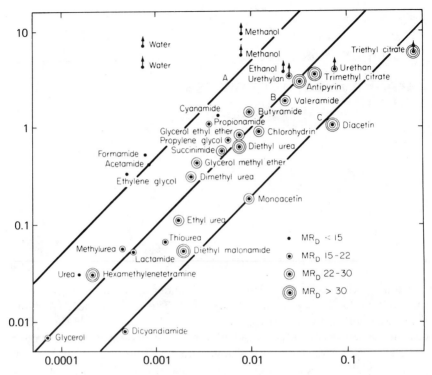

Fig. 1.1 The permeability of cells of *C. ceratophylla* to organic nonelectrolytes of different oil solubility and different molecular size. Ordinate: $PM^{1/2}$ (P in cm h^{-1}). Abscissa: olive oil–water partition coefficients. MR_0 is the molar refraction of the molecules depicted, a parameter proportional to the molecular volume. [Reproduced with kind permission of Stein 1967].

slowly to be caused by a simple change in surface tension. Her experiments suggested the presence of a discrete, elastic membrane that was distinctly more elastic than the material in the interior of the cell.

Such experiments indicated the presence of a discrete membrane composed of organic molecules. However, they said nothing of the size and properties of the membrane. To gather such information, the molecules that constitute the membrane must be examined.

2. MEMBRANE MOLECULES

Although actual membranes may contain an extremely large number of different molecules, some types of molecules are generally present in all membranes. Because such molecules constitute the bulk of the membrane medium, their structures can give some insights into the nature of the membrane itself.

2.1. LIPIDS

Many lipid molecules are formed from the glycerol molecule.

$$
\begin{array}{ccc}
\text{H} & \text{H} & \text{H} \\
| & | & | \\
\text{H} & \text{O} & \text{O} \\
| & | & | \\
\text{H}-\text{C}-\text{C}-\text{C}-\text{H} \\
| & | & | \\
\text{H} & \text{H} & \text{H}
\end{array}
$$

The three alcohol groups can form esters with fatty acids, and two of them are esterified with long chains (18–26 carbon atoms) in lipid molecules. The third OH is bound to a phosphate group, and this leads to a lipid molecule with the structure

$$
\begin{array}{cc}
\text{R} & \text{R}' \\
| & | \\
\text{O}=\text{C} & \text{C}=\text{O} \\
| & | \\
\text{O} & \text{O} \\
| & | \\
\text{H}-\text{C}-\text{C}-\text{C} \quad \text{O} \\
| \quad | \quad | \quad | \\
\text{H} \quad \text{H} \quad \text{O}-\text{P}-\text{O}-\text{X} \\
| \\
\text{O}
\end{array}
$$

where R and R′ are the long hydrocarbon chains of the fatty acid molecules. The X group is used to characterize the different lipid molecules. If X = H, the entire molecule is called phosphatidic acid, although there are still a large number of actual molecules that satisfy this definition because R and R′ are not specified.

In many cases, the X group will also have an associated charge. A number of such groups have a net positive charge that complements the negative charge associated with the phosphate group. For example, phosphatidyl choline is formed by esterifying phosphatidic acid with choline,

$$X = CH_2CH_2N^+(CH_3)_3$$

Phosphatidylethanolamine is formed when

$$X = CH_2CH_2NH_2$$

and phosphatidylserine is formed when

$$
X = H_2C-\overset{\displaystyle \text{H}}{\underset{\displaystyle \text{NH}_2}{\text{C}}}-COOH
$$

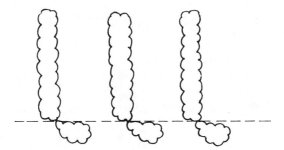

Fig. 1.2 A schematic model of a polar lipid. The vertical portions represent the hydrocarbon chains.

All three molecular species have a nitrogen group, which may have a proton or may acquire a proton in solution. The total lipid may then be divided into two basic regions. The hydrocarbons R and R′ constitute a hydrocarbon region of the lipid molecule, while the region containing the phosphate and the X group is intrinsically charged and thus tends to be polar. Molecules such as this, containing distinct regions of nonpolar (hydrocarbon) and polar regions, are called amphipathic molecules. Note that only the polar region dissolves in water.

In many of the models that will be discussed, the detailed molecular structure of the lipid molecules need not be known explicitly. However, such models make use of the essential character of distinct polar and nonpolar regions. Thus, the molecule will often be illustrated as in Fig. 1.2, where the vertical portions signify the hydrocarbon chains and the polar region contains the charged moieties.

2.2. STEROIDS

The steroids are fused ring systems. When found in membranes, they exert a stabilizing influence on the overall membrane structure. For example, the cholesterol molecule has the structural form

Cholesterol

Although the structure appears to be quite different from that of the lipids because of the fused ring system, a closer inspection of the cholesterol molecule suggests that it is similar to a long hydrocarbon chain. The cholesterol molecule can be illustrated schematically as

Such observations have led some workers to suggest that cholesterol molecules can form a 1:1 complex with lipids. The complex is illustrated in Fig. 1.3.

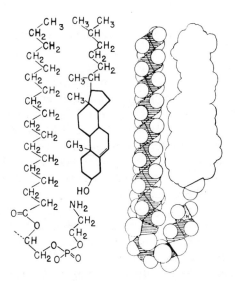

Fig. 1.3 A possible arrangement for the complex between cholesterol and phospholipid in myelin. [Reproduced with kind permission of Lucy 1965.]

2.3. PROTEINS

Proteins are polymers produced by linking amino acids. The amino acids have the general form

$$R-C-COO^-$$
$$| $$
$$NH_3^+$$

The linkage uses peptide bonds to give a structure of the form

$$\begin{matrix} O & H & R & O & H & R' \\ \| & | & | & \| & | & | \\ -C- & N- & C- & C- & N- & C- \end{matrix}$$

The different amino acids in the chain can be used to produce a large variety of different proteins characterized by different arrangements of amino acids along the chain. For 20 different amino acids in a chain of length N, there are 20^N different possible proteins. For example, any one of the 20 amino acids may appear as the first link in the polymer. Likewise, any of the 20 may appear in the second position. Using just these two positions, there are already $20 \times 20 = 400$ different arrangements.

3. SOME SURFACE PROPERTIES OF LIPIDS

The existence of distinct polar and nonpolar regions in lipids generates some interesting properties when the lipid is placed on the surface of a polar medium such as water. The hydrophilic polar heads can dissolve in water, while the hydrophobic nonpolar groups are insoluble. The polar groups "root" the hydrocarbon "tree" to the water surface.

When the lipid is added to the surface, the hydrocarbon chains show no attraction for each other and will therefore tend to move apart to minimize interactions. No repulsive interactions are involved; the molecules simply spread over the water surface to minimize interference with each other. The two-dimensional surface is very similar to a three-dimensional gas. The gas molecules expand to fill the entire container.

The molecules on the surface can be compressed if barriers are placed on the surface. Such barriers can then be moved closer together to compress the surface film. However, the molecules can only be compressed a certain amount before they begin to interfere with each other. If this interference becomes too large, the molecules may slip under these barriers or may buckle over each other to form multilayers. To prevent this, the surface pressure must be monitored in some way.

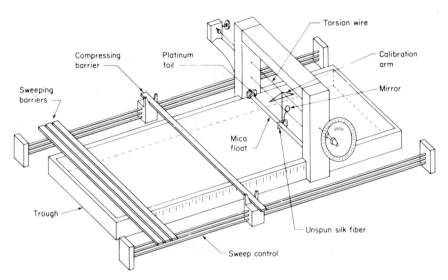

Fig. 1.4 A modern form of a Langmuir film balance. [Reproduced with kind permission of Moore, 1972.]

One device for monitoring the surface pressure is the Langmuir trough, which is shown in Fig. 1.4. A trough is filled with water and the surface is swept clean with a movable barrier. In order to compress the film, there must be two barriers. A movable surface barrier is moved along the edges of the trough to change the total surface area available to the molecules deposited on the surface. At the other end of the trough, a surface barrier of mica is connected to the sides of the trough with flexible platinum strips to permit it to move. The float is suspended to a torsion wire. If the barrier is moved by the increase in molecular surface pressure, this wire will twist. The force necessary to return the wire to its original position is a direct measure of the surface pressure. The magnitude of this force can be calibrated for the torsion wire by attached weights of known value, suspended from an arm attached to the wire.

To use the balance, a known amount of the lipid is added to the surface with some solvent; the solvent immediately evaporates, leaving the surface material. The surface barrier is then moved to decrease the surface area, and the force needed to balance the resultant surface pressure is recorded for each area. During the compressions for large area, the lipids encounter little resistance (surface pressure), because they are still separated. As the film compresses further, however, the surface pressure begins to rise quite rapidly. In this region, the molecules interfere markedly and the hydrocarbon chains are forced perpendicular to the surface. The plot of surface

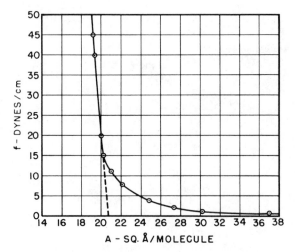

Fig. 1.5 Isotherm of f (surface pressure) versus A (area) at 20°C. Stearic acid on distilled water. [Reproduced with kind permission of Moore, 1957].

pressure versus surface area is shown in Fig. 1.5 and the surface arrangements for both the uncompressed and compressed monolayers are shown in Fig. 1.6. Since the compressed monolayer has the chains in a vertical position, it displays a considerable amount of organization and can be used as a simple model for a membrane.

The shape of the pressure-versus-area curve of Fig. 1.5 is extremely similar to the pressure–volume isotherm observed for gases and is, in fact, a two-dimensional analog of the gaseous system.

The pressure-versus-area curve enables us to estimate the area occupied by each molecule on the surface. Since a known number of molecules have been added to the surface, the area per molecule is easily calculated. However, the total area of the surface must be determined. The usual procedure involves an estimate of the surface area of the compressed surface layer for zero surface pressure. Obviously, the film would spread without this surface pressure. However, the "compressed" region of the pressure–area isotherm can be

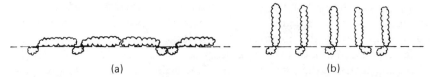

Fig. 1.6 Organization of lipids in a surface monolayer. (a) Compressed monolayer. (b) Uncompressed monolayer.

extrapolated to zero pressure, as shown in Fig. 1.5. The intersection of this extrapolated line with the area axis is generally chosen as the surface area of the compressed membrane.

The area per molecule A_{mol} is calculated from the total area A_{tot}, and the total number of molecules N on the surface,

$$N = nN_0 \tag{1.1}$$

where N_0 = Avogadro's number, as

$$A_{\text{mol}} = A_{\text{tot}}/N \tag{1.2}$$

Since the total number of molecules added is known, the volume per molecule can be calculated if the density of the surface material is known. The molar density is

$$\rho_{\text{m}} = \rho/M \tag{1.3}$$

where ρ_{m} is the molar density, ρ is the density, and M is the molecular weight. The inverse of the molar density gives the volume per mole of the material, and this quantity is then divided by Avogadro's number to determine the volume per molecule. When both the volume and area per molecule are known, the height of the monolayer can be estimated, because

$$h = (\text{volume/molecule})/(\text{area/molecule}) \tag{1.4}$$

Consider the case in which 0.063 mg stearic acid is dispersed on the water surface. Extrapolation in the compressed region back to zero surface pressure yields a total area of 300 cm². Since the molecular weight of stearic acid is 284 g/mole, the area per molecule is

$$\begin{aligned} A_{\text{mol}} &= (300 \text{ cm}^2)[(284 \text{ g mol}^{-1})/(6.3 \times 10^{-5} \text{ g})] \\ &\quad \times [1 \text{ mole}/(6.023 \times 10^{23} \text{ molecules})] \\ &= 22.5 \times 10^{-16} \text{ cm}^2/\text{molecule} = 22.5 \text{ Å}^2/\text{molecule} \end{aligned} \tag{1.5}$$

The volume per stearic acid molecule is determined from the density of 0.85 g/cm³ for this molecule,

$$\begin{aligned} V/\text{molecule} &= (1 \text{ cm}^3/0.85 \text{ g})(284 \text{ g}/1 \text{ mole})(1 \text{ mole}/6.02 \times 10^{23} \text{ molecule}) \\ &= 562.5 \times 10^{-24} \text{ cm}^3/\text{molecule} \\ &= 562.5 \text{ Å}^3/\text{molecule} \end{aligned} \tag{1.6}$$

The height of the monolayer is then

$$(562.5 \text{ Å}^3/\text{molecule})/(22.5 \text{ Å}^2/\text{molecule}) = 25 \text{ Å}/\text{molecule} \tag{1.7}$$

Although the monolayer is fully compressed in the experiments, the surface pressure might still be increased. However, at these large surface pressures, the character of the monolayer is lost. The surface may "buckle,"

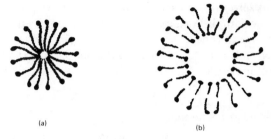

(a)

(b)

Fig. 1.7   Micellar and vesicular lipid organization. (a) Micelles. (b) Vesicles.

forcing the lipid molecules below the water surface. The hydrocarbon chains will adjust to minimize their interaction with the water. They will do so by forming associated globules, or micelles, in which the polar groups of the lipids are all located on the perimeter of the sphere while the hydrocarbon chains remain inside.

Vesicles can also be formed from lipids. The vesicle is a lipid bilayer that surrounds an internal aqueous solution. The polar groups of one lipid layer face the outer solution, while the polar groups of the inner layer face the internal solution. The micellar and vesicle systems are shown in Fig. 1.7. Both can be used for the study of organized membrane systems.

4. THE GORTER–GRENDEL EXPERIMENT

By the 1920s, the existence of some type of organic barrier between the cells and their environment was well established. However, the nature of this barrier was still unknown. Erythrocyte (red blood) cells could be lysed and their contents could be removed, leaving the cell membrane a "red cell ghost." Because these lysed red cell membranes contained lipid molecules, Gorter and Grendel (1925) performed a quantitative experiment to determine the nature of lipid organization in the membrane, that is, whether lipids formed a bilayer membrane.

Gorter and Grendel extracted the lipid from a known number of such red cell ghosts. This lipid was then spread on a surface of a Langmuir trough and the total area of lipid occupied by the lipid in its "compressed" form was determined. Say, for example, that 60 cm^2 of surface was occupied by these lipids. Now, if these lipids formed the membranes, it could easily be verified by measuring the total surface area associated with one cell. When this area was multiplied by the total number of cells present in the original extracted set of ghosts, that surface area should match the area observed on the Langmuir trough. When such calculations were completed, the total cell surface area that had extracted for a 60-cm^2 monolayer was found to be only 30 cm^2—that is, the 30-cm^2 surface area of cells had produced a monolayer of lipid that occupied twice the area. The membrane around the red blood cells had to be a bilayer.

Because the bilayer model was reasonable, the data of Gorter and Grendel was accepted quite readily. However, in 1966, Bar, Deamer and Cornwell repeated the experiments of Gorter and Grendel. Although the lipid extraction of Gorter and Grendel was found to be incomplete, these researchers also determined that the original cell surface-area calculations of Gorter and Grendel had also been in error; two errors cancelled to produce the ratio of 2.

5. SURFACE TENSION

The properties of molecules at interfaces tend to differ from the properties of these same molecules buried deeply in the bulk phase. In the bulk phase, the molecules are completely surrounded by other molecules of the same phase. If there are attractions between such molecules, such attractions come from all directions because the molecules are distributed homogeneously in the solution. At the interface, however, these same molecules interact with the molecules, but these molecules are no longer distributed homogeneously. The molecules at the surface are attracted by the bulk-phase molecules in one direction, while experiencing different attractions from molecules in the second phase bounding the interface. Qualitatively, such an attractive system would shrink to the smallest possible surface area to minimize interactions with the neighboring phase. The energy required to increase or decrease the surface area is a function of the molecules that constitute the phase.

Because energy is required to change the area of an interface, this energy must be proportional to the change in area. If the change in surface area is $d\sigma$, then the corresponding change in energy U is

$$dU = \gamma \, d\sigma \tag{1.8}$$

The surface tension, γ, is the proportionality factor that relates energy and surface area. For consistency, γ must have the units of energy per unit area, for example, joules per square meter. The surface tension is also expressed in the equivalent units of force per unit distance—for example, newtons per meter—because the energy is equivalent to force times distance, so that

$$J/m^2 = (N \cdot m)/m^2 = N/m \tag{1.9}$$

The units of force per unit distance reflect one of the techniques used to measure surface tension. Because the molecules at the surface experience interactions different from those in the bulk phase, different forces should be required to move an object through the interface. The du Noüy tensiometer measures the force required to pull a circular ring through the interfacial region. The circumference of the ring defines the length over which this force acts. Because both the top and bottom of the ring must be pulled through the interface, the actual length is twice the circumference. For a ring of radius r, the force is then

$$\text{force} = 4\pi r \gamma \qquad (1.10)$$

Since r is known and the force can be measured, the surface tension can be determined experimentally. From an energetic point of view, the tensiometer determines the energy necessary to alter the interfacial molecules so that the ring can pass through the interface.

6. SURFACE TENSION MEASUREMENTS ON MEMBRANE SYSTEMS

Because cell membranes represent the interface between the molecules of the cell interior and the external bathing solution, a measurement of this interfacial surface tension would be invaluable. By comparing the cell surface tension with those observed for different molecular phases and water, the composition of the cell membrane could be determined. For example, water–hydrocarbon interfaces exhibit surface tensions of 0.05 N/m or 50 mN/m. If the cell–water interface is similar to the water–hydrocarbon boundary, then similar magnitudes should be observed for the cell surface tension.

Although this idea is simple in principle, it is a difficult experimental problem. A large device such as the du Noüy ring cannot be pulled through the cell membrane. Alternative types of forces must be applied to the cell while concomitant changes in its surface area are observed.

One technique for the determination of cell surface tension takes advantage of the fact that there are regions of different density within the cell. For example, some fish eggs contain a nucleus that is denser than water and an oil phase that is less dense than water. The nucleus would then fall relative to water in a gravitational field, while the oil would be buoyed relative to the water. Harvey and his co-workers used a centrifuge system to amplify the gravitational forces experienced by the cell materials. The denser material

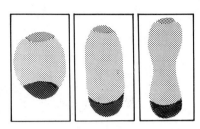

Fig. 1.8 Three stages in the pulling by centrifugal force of an unfertilized *Arbacia punctulata* egg into two half eggs: oil at top, yolk and pigment at bottom. Depicted immediately on stopping centrifuge.

within the cell was thrown outward with respect to the less dense material producing an elongation of the cell. When the length of the distorted cell exceeds the cylindrical circumference, the cell becomes unstable and splits into two parts, as shown in Fig. 1.8. The force at the instant of rupture must equal the surface tension times the circumference of the cylinder at the rupture point. This force must equal the difference in gravitational force experienced by the denser and less dense cell regions. The gravitational acceleration in the centrifuge is the angular frequency ω of the centrifuge multiplied by the radial distance from the rotation axis to the cell, r'. The masses of the two materials relative to water are

$$m_H = V_H(\rho_H - \rho_S) \tag{1.11}$$

$$m_L = V_L(\rho_L - \rho_S) \tag{1.12}$$

where ρ_H and ρ_L are the densities of the nucleus and the oil and ρ_S is the density of the solution, and V_H and V_L are the volumes of the nuclear and oil regions, respectively. The net force on the cell is then related to the difference of these two masses,

$$\omega^2 r'(m_H - m_L) = 2\pi r \gamma \tag{1.13}$$

By photographing the cells at the instant of break, the relevant parameters could be determined and the surface tension could be established.

Cole (1932) distorted the cell by placing it under a microbeam with a weight of approximately 1 μg. By measuring cell radii for each side of the cell under the beam (Fig. 1.9) and the flattened area A of the cell under the microbeam, the cell surface tension could be determined from the relationship

$$mg/A = \gamma(1/r_1 + 1/r_2) \tag{1.14}$$

where the mass m of the microbeam determines the pressure on the cell.

The values of surface tension observed for cell membranes were significantly smaller than those observed for a hydrocarbon–water interface (50 mN/m). For example, mackerel eggs yielded surface tension of 0.6 mN/m. Unfertilized *Arbacia* (sea urchin) eggs gave 0.2 mN/m. These results indicated

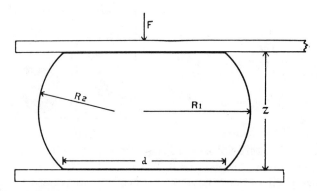

Fig. 1.9   Cole microbalance used to measure cell surface tension. [Reproduced with kind permission of Cole (1932).]

that the membrane interface was not a simple hydrocarbon–water interface. Since protein molecules could reduce the surface tension, the observed low surface tensions were attributed to the presence of protein molecules in the surface region.

Surface tensions can also be measured by distorting lipid bilayers where the membrane composition is well known and protein can be completely excluded. For these systems, the surface tension is also below 10 mN/m, even though no protein is present. This is due to the small size of the membrane. The bilayer experiences forces from both aqueous solutions, and this symmetrical attraction experienced by the bilayer molecules reduces their net surface tension.

7. X-RAY SCATTERING BY MEMBRANES

Light rays are reflected by a polished surface because the wavelength of light is long relative to the irregularities of the surface. If the surface is lined with a regular series of grooves—that is, a grating—then different light wavelengths are reflected in different directions. The light wavelengths are dispersed by the regular geometry of the grating.

Crystals of atoms or molecules will have a very regular structure in the 1–20-Å range. This regular structure could be probed by observing the dispersive properties of electromagnetic radiation in the angstrom range, that is, x rays.

The crystal structure can be visualized as a set of parallel planes with angstrom separations. Waves reflected off the upper plane will obviously

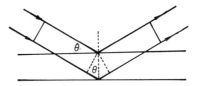

Fig. 1.10 Reflections from two parallel planes of atoms when the angle of incidence of the x rays is θ; d is the distance between planes.

have a phase different from that of the waves reflected off the plane below. If the distance to the lower plane is one full phase of the incoming radiation, then both reflected waves will interfere constructively. When the two planes are separated by a distance d, x rays that are reflected or scattered from the lower plane will travel an additional distance dictated by the angle at which the ray strikes the surface. This is illustrated in Fig. 1.10. The angle of the incident radiation relative to the crystal plane is θ. If the incident ray and the crystal surface are both rotated by $90°$, it is easy to show that the angle θ also opens onto the extra distance that the beam must travel to reach the lower plane. Both the incident ray and the reflected ray will travel $d \sin \theta$ relative to a ray reflected off the upper plane. The net distance will then be $2d \sin \theta$, and this must be equal to an integral number of x-ray wavelengths λ if the two rays are to interfere constructively. The Bragg relation is then

$$n\lambda = 2d \sin \theta, \qquad n = 1, 2, \ldots \qquad (1.15)$$

and the scattering angles θ that lead to constructive interference become the experimental parameters of interest.

The angle θ must be defined in terms of the experimental system. A powder that consists of crystals at all orientations is placed in the path of a beam of x rays, as shown in Fig. 1.11. If one of these crystals is oriented at the proper angle θ with respect to the incoming beam, constructive scattering will take place. If a film strip is wrapped around the sample, the film will be exposed at certain angles relative to the source. If the angle between the source and scattered beam is Φ as shown in the figure, then

$$\Phi = \pi - 2\theta$$

The geometry is illustrated in Fig. 1.11.

When nerve membrane, which is noncrystalline, is placed into an x ray beam, a relatively sharp line is observed at a larger θ corresponding to a smaller interplane distance. The scattering angle corresponds to an inter-plane distance d of 4.7 Å. In our discussion of surface pressure, we deduced a surface area of 22.5 Å^2 for the stearic acid molecule. This suggests a mean distance between stearic acid molecule centers of 4.7 Å. The phospholipids of a membrane system would have two fatty acid chains per molecule.

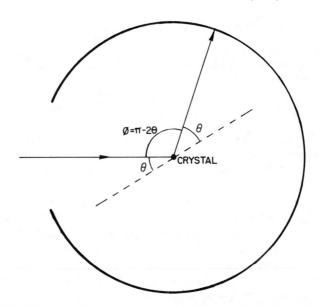

Fig. 1.11 Schematic of an x-ray system showing the relationship between the observed scattering angle Φ and the angle of incidence θ. The dotted line illustrates the orientation of the crystal face that is reflecting the x rays.

The x-ray studies suggest that all the chains are roughly parallel to each other and have a mean separation of 4.7 Å. The membrane has an ordered "liquid crystalline" structure in a direction perpendicular to the membrane surface.

8. BIREFRINGENCE MEASUREMENTS

The index of refraction is defined as the ratio of the speed of light in a vacuum to the speed of light in the medium of interest,

$$n = c_{\text{vacuum}}/v_{\text{medium}} \tag{1.17}$$

Although this relationship implies that the medium will have only a single index of refraction, this need not be the case. A structured medium (such as a membrane) may have different indices of refraction along different spatial axes in the membrane. If this does occur, the system will appear to rotate a polarized beam of light that passes through it. This is the phenomenon of birefringence.

The ability of a given molecule to reduce the speed of light is proportional to the magnitude of the permanent dipoles or the induced dipoles which can be generated, This magnitude is reflected in the dielectric constant D of the medium which is discussed in more detail in Chapter 3. The electric permittivity in the medium, ε, is related to permittivity ε_0 in a vacuum via the relation

$$\varepsilon = \varepsilon_r \varepsilon_0 \qquad (1.18)$$

where ε_r is a relative constant: for example, $\varepsilon_r(H_2O) = 80$.

A light wave is a transverse wave: that is, the amplitude of the wave varies in a direction perpendicular to the direction in which the wave is moving. For a beam of light moving in the z direction, oscillations of the electric field are possible in any direction in the x–y plane. However, these oscillations can always be resolved into projections on the x and y axes, E_x and E_y. If the light beam can be directed through a polaroid sheet with preferential absorption of one of these components, the resultant light beam will contain only the remaining projection. The resultant beam is then plane-polarized or linearly polarized. Other types of polarization are possible. For example, the electric field vector might spiral as it moves through space. This would be circularly polarized light.

If a beam of light is passed through two plane polarizers oriented at right angles to each other, no light will pass through, because one set of components is eliminated by the first polarizer and the remainder are eliminated by the second polarizer. However, if a myelin membrane is inserted between the two polarizers, light will be transmitted through these two perpendicular polarizers. The nerve membrane rotates the vector of the plane polarized light so that a component is transmitted through the second polarizer.

This ability of an ordered system to rotate plane-polarized light is called birefringence, and it arises in systems where different spatial orientations of the system have different indices of refraction. For the myelin membrane, the ordered hydrocarbon chains of the membrane present a specific index of refraction to an electric vector oriented parallel to the chain direction and a second index of refraction to an electric vector perpendicular to these chains. In a disordered system, the chains face in all directions and an average index of refraction is observed. An ordered arrangement is required to rotate the light beam.

To simplify the analysis, assume that the axes of the membrane system are oriented at 45° to the two electric field vectors determined by the orientation of the polarizers. The polarized electric vector that leaves the first polarizer will then appear to lie midway between the x and y axes of the membrane, as shown in Fig. 1.12(a). The electric vector will then project equal components

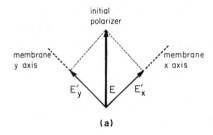

(a)

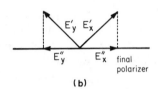

(b)

Fig. 1.12 Orientation of major membrane axes relative to the electric field vector of plane polarized light. (a) Equal projections of polarized light onto the major membrane axes. (b) Projections of electric field components of light that are transmitted by the second polarizer.

onto the perpendicular x and y axes. For the 45° angle, these components will have the values

$$E_x = E/\sqrt{2} \quad \text{and} \quad E_y = E/\sqrt{2} \tag{1.19}$$

Note that the actual intensity of light in each direction will be proportional to the square of the wave amplitude. When the amplitudes are squared, the resultant intensity of the incoming beam is divided equally between the two membrane axes.

Although the electric field vectors are now directed to specific axes, their amplitude does not remain constant. E_x and E_y are maximal amplitudes for oscillating electric waves of the form

$$E_x(t) = E_x \sin(\omega t - \delta) \tag{1.20}$$

where $\omega = 2\pi\nu$ is the angular frequency of the light and δ is a phase delay.

The rotation of the plane-polarized light is caused by the different velocities of light for the x and y components of the electric vector as they pass through the membrane system. If the wave moves more slowly for the component directed along the y direction, then this wave will have a phase delay δ relative to the wave along the x direction. The two electric field vectors that leave the membrane are then

$$E'_x(t) = E_x \sin \omega t \tag{1.21}$$

$$E'_y(t) = E_y \sin(\omega t - \delta) \tag{1.22}$$

In order to determine if any light will pass through the remaining cross-polarizer, these two components must now be projected onto the horizontal axis of the second polarizer, because this is the orientation which will transmit light. The projections are shown in Fig. 1.12(b). The E'_x component leaving the membrane projects onto this horizontal axis is

$$E''_x = E'_x/\sqrt{2} = (E/2)\sin(\omega t) \qquad (1.23)$$

where E''_x has been related to the original light amplitude, E. The y component is

$$E''_y = -E'_y/\sqrt{2} = -(E/2)[\sin(\omega t - \delta)] \qquad (1.24)$$

where the minus sign reflects the projection onto the negative x axis as shown in Fig. 1.12(b). The net electric vector that leaves the second polarizer is then the vector sum of these two components,

$$E_o = E''_x + E''_y = (E/2)[\sin(\omega t) - \sin(\omega t - \delta)] \qquad (1.25)$$

The expression can be simplified with the trigonometric identity,

$$\sin x - \sin y = 2\sin[(x - y)/2]\cos[(x + y)/2] \qquad (1.26)$$

When $x = \omega t$ and $y = \omega t - \delta$, the expression for E_0 reduces to the form

$$E_o = (E/2)(2)\sin(\delta/2)\cos[(2\omega t - \delta)/2] \qquad (1.27)$$
$$= E\sin[\delta/2]\cos[\omega t - (\delta/2)] \qquad (1.28)$$

When the membrane is oriented at 45° to the crossed polarizers and the thickness of the membrane region is sufficient to produce the 180° shift, the equation reduces to

$$E_o = E\sin(90°)\cos(\omega t - 90°) = E\sin(\omega t) \qquad (1.29)$$

For this special condition, the original light beam will traverse both polarizers with no loss of amplitude. The light will have rotated through 90°.

Schmidt (1936) used birefringence observations to indicate ordered structure in frog nerve myelin. This spatial birefringence could be eliminated by bathing the axonal system in alcohol to destroy the two-dimensional structure of the myelin system.

PROBLEMS

1. In the Gorter and Grendel experiments with the Langmuir surface balance, the surface area calculated for 1 g of red blood cells was 0.47 m². On the surface balance, 50 mg of the extracted lipid completely covered a surface of 0.047 m². Are these results consistent with a bilayer model? Could the molecular weight of the lipid be determined?

2. One gram of stearic acid (MW = 284 g mole^{-1}) is spread on a large surface, for example, a lake. If the surface area of the compressed molecules is 22 Å²/molecule, determine the surface area covered by the stearic acid.

3.   Using the information in Problem 2 and the height h of the stearic acid chain, estimate the surface area occupied by the uncompressed molecules. Assume the molecules will spread until they lose contact with each other.

4.   Discuss the difference between the x-ray scattering angle θ and the phase delay Φ for scattering from a crystalline array.

5.   In a crystal of the molecule AB, the x rays are directed at a face perpendicular to the inter-atomic axis of the molecule. If the separation between centers of the AB molecules along the internuclear axis is 10 Å and the AB bond distance is 2.5 Å, calculate the phase delay for planes of B atoms relative to planes of A atoms.

6.   KCl forms a simple cubic unit cell with K^+ and Cl^- ions on adjacent corners. Determine the density of the KCl crystal if the first reflection is observed at $\theta = 14°12'$.

7.   In a birefringence experiment, find an expression for the light amplitude transmitted by a pair of crossed polarizers oriented at 30° relative to the molecular axes.

8.   The stearic acid molecule has a net height of 25 Å/molecule. Use this information and the fact that the C—C bonds are tetrahedral to estimate the carbon–carbon bond distance. How does your result compare with the known distance?

Chapter 2

Thermodynamics

1. THE FIRST LAW OF THERMODYNAMICS

The first law of thermodynamics is a quantitative expression of two important laws of physics: (1) energy is conserved, and (2) heat and work are both forms of energy. The first law simply provides the formalism for expressing these facts.

Consider a sealed container with heat-conducting sides at some temperature T_1. If this container is now lowered into a water bath at some higher temperature T_2, an amount of heat Q will flow across the boundary into the container. The heat flow will cease when the temperature of the container and bath become equal.

Consider what has now happened at the molecular level. The heat must be contained in the increased translations and internal motions of the molecules in the sealed container—that is, the heat has gone to increase the internal energy of the molecules within the container. Thus, a change in internal energy ΔU is produced by the addition of the heat Q,

$$\Delta U = Q \tag{2.1}$$

If a movable piston is now placed on one side of the container so that the container volume V can change, work can be done by the molecules within the container. Because energy is conserved, any work must be done at the expense of the internal energy of the molecules if there is no additional heat flow into the container. The work W is labeled positive when the piston moves outward against a pressure p to increase the container volume. Because this lowers the energy, the internal energy is

$$\Delta U = Q - W \tag{2.2}$$

21

Any alternative forms of work must obtain their energies from the molecules of the container as well, so the first law is often expressed in terms of both PV (pressure–volume) work and other forms of work,

$$\Delta U = Q - W_{PV} - W_{other} \tag{2.3}$$

For infinitesimal changes in work and heat added, the first law can be expressed in differential form,

$$dU = dQ - dW \tag{2.4}$$

For systems with only PV work, the actual work done is determined by the magnitude of the external pressure. This pressure, or its related force F,

$$F = P_{ex} A_p \tag{2.5}$$

where A_p is the area of the piston face, is the parameter that would actually move a wheel or external object; it is the connection between the molecules of the sealed container and work in the outside world. If the volume of the container can change, given this external pressure, work will be produced and can be expressed as

$$dW = P_{ex} \, dV \tag{2.6}$$

There are obviously a large number of different ways in which we can increase or decrease the internal energy of the system. However, there are a limited number of ways to store energy in the molecules, and this stored energy can come from either work or heat. For this reason, it is convenient to focus on the internal energy. Rather than focusing on the entry of heat or work, we examine parameters of the system itself that might signal changes in the internal energy. For example, changes might be expected if the temperature or volume of the system changes. This can be expressed for differential changes in the internal energy as

$$dU = (\partial U/\partial T)_V \, dT + (\partial U/\partial V)_T \, dV \tag{2.7}$$

This differential expresses the important state property of the internal energy. During most changes in a system, both the volume and the temperature might be changing simultaneously. Equation (2.7) expresses the fact that we can calculate this internal energy by selecting an alternative path. The temperature is first changed from its initial to its final value while the volume is held constant. Then the temperature is held constant until the final volume is reached. The internal energy for the transition $U(T_i, V_i) \rightarrow U(T_f, V_f)$ done in this two-step sequence is identical to the energy that would be observed if both variables were varied simultaneously.

The partial derivative $(\partial U/\partial T)_V$ is defined as the heat capacity at constant volume, C_V. The derivative $(\partial U/\partial V)_T$ is zero when intermolecular attractions are zero. The derivative itself is often confusing because it is sometimes associated with external work. For instance, if the volume changes, then work is being done at the expense of the internal energy, so this energy must change. However, it is necessary to realize that this derivative is associated with the system alone. For example, a gas could be placed in an insulated container with no heat flow ($Q = 0$) and then expanded against a vacuum so that $P_{ex} = W = 0$, as shown in Fig. 2.1. Under such circumstances, the temperature of the gas would fall slightly. Some of the energy of the molecules would be used to separate the molecules as they expanded to the larger volume. The energy is used to overcome the attractions between the molecules.

Both Q and W have been expressed in terms of external parameters, such as P_{ex}. It is sometimes convenient to express the first law in terms of internal pressures, temperatures, and so on. We can do this by bringing the system to the special limit of reversible behavior. Consider the work term as an example. If the external pressure is much lower than the pressure of the system, the piston will move out spontaneously. To restore it to its original position, work will be required. As the internal and external pressures approach each other, it will take less work to restore the piston. No work will be done or required for restoration when $P_{int} = P_{ext}$. However, we can produce work by lowering the external pressure by an infinitesimal amount so that the volume changes infinitesimally. By repeating this step over and over, we will eventually reach the final volume of interest. Throughout this entire expansion process, $P_{int} = P_{ext} - \delta P$. Because δP is small relative to P, P_{int} can be substituted for P_{ext} in the work expression such that

$$dW = P_{ex}\, dV = P_{int}\, dV \qquad (2.8)$$

In the reversible limit, the pressure of the system can be used in the first law expression.

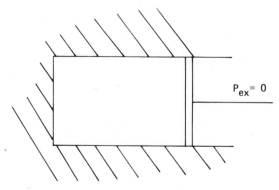

Fig. 2.1 Gas in an insulated container expanding against zero external pressure: $Q = 0$, $W = 0$.

2. ENTROPY

A system performs work only if P_{ex} and P_{int} differ. In the reversible limit, the two pressures are allowed to approach each other so that the work takes place in a series of small steps. At any point, P_{ex} could be increased infinitesimally and the piston would move in the opposite direction.

A similar situation arises when we consider the flow of heat. To force heat into the system, the external temperature must be higher than that of the system. As this temperature difference becomes larger, it requires more work to remove this energy from the system. By analogy with the pressure-volume example, T_{ex} will be the parameter that determines whether heat flows into or out of the system. In order to bring heat flow to a reversible limit, the external and internal temperatures must differ by an infinitesimal amount, so that $T_{ex} = T_{int}$. At this point, heat may be transferred in either direction by a slight change in the external temperature. The reversible heat and the internal temperature, T, are now related through a proportionality factor called the entropy S:

$$dQ_{rev} = T\,dS \tag{2.9}$$

or

$$dS = dQ_{rev}/T \tag{2.10}$$

From Eqs. (2.4), (2.8), and (2.9), the first law can now be expressed in terms of variables for the system,

$$dU = T\,dS - P\,dV \tag{2.11}$$

In addition to relating the internal temperature and reversible heat of the system, the entropy has additional properties that make it a special function. For example, consider the case of an ideal gas that is expanded reversibly from V_1 to V_2 at a constant temperature. For an ideal gas, the internal energy cannot change if the temperature is constant, so

$$dU = 0 \tag{2.12}$$

and

$$0 = dQ_{rev} - dW_{rev} = dQ_{rev} - P\,dV \tag{2.13}$$

Because

$$P_{int} = RT/V \tag{2.14}$$

then

$$dQ_{rev} = dW_{rev} = [RT/V]\,dV \tag{2.15}$$

Integrating between V_1 and V_2 gives

$$Q_{rev} = RT\ln[V_2/V_1] \tag{2.16}$$

Substituting Eq. (2.16) into the entropy expression gives

$$\Delta S = RT\ln[V_2/V_1]/T = R\ln[V_2/V_1] \tag{2.17}$$

The change in entropy in this case is completely independent of the temperature. The entropy will increase if the volume available to the gas is increased. If the volume of the gas is doubled in an expansion, each gas molecule can traverse twice as many locations; it therefore becomes more difficult to locate this molecule. The entropy is a measure of the chaos that results during the expansion of the gas.

To illustrate this point in a second way, consider the entropy change when the volume is doubled,

$$\Delta S = R\ln(2) = kN_0\ln(2) = k\ln(2^N) \tag{2.18}$$

where k is Boltzmann's constant and N_0 is Avogadro's number. The factor 2^{N_0} is the expression of particular interest. If the volume of the cell is doubled, then the number n of molecular-sized cells within the volume would also be doubled to $2n$. The probability of finding a molecule in a particular cell is now halved, because the molecule is free to move to all cells in the volume. Alternatively, the number of cells in which it can be located is now doubled. There are now four times as many arrangements for the two molecules in the expanded volume, as illustrated in Fig. 2.2. For N_0 molecules, the total increase of arrangements when the volume is doubled is 2^{N_0}. In general, when the number of possible molecular arrangements increases from Ω_1 to Ω_2, the increase in entropy will be

$$\Delta S = k\ln[\Omega_2/\Omega_1] \tag{2.19}$$

Equation (2.19) carries the implicit assumption that each of the molecular cells is equally likely to accept a molecule. The probability of one specific arrangement is

$$p = 1/\Omega \tag{2.20}$$

and the entropy associated with this system is

$$S = k\ln(\Omega) = -k\ln(p) \tag{2.21}$$

This expression can be generalized for situations in which all configurations are not equally probable. If the probability for configuration j is p_j, the entropy can be written as a weighted sum of the logarithms of these probabilities,

$$S = -k\sum p_j \ln p_j \tag{2.22}$$

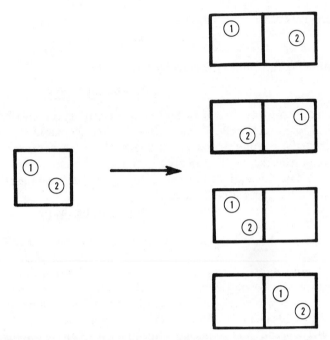

Fig. 2.2 The increased configurations possible for two particles when the volume of a container is doubled.

To illustrate the use of this equation, consider a DNA molecule that contains four possible nucleotides. If the molecule were completely random, we would expect to find equal fractions (0.25) for each of the four molecules C, G, A, and T. The entropy for such a random sequence is then

$$S = \sum_4 0.25k \ln(0.25) = k \ln(0.25) = 1.4k \tag{2.23}$$

As the DNA molecule evolved to produce specific types of proteins, the purely random distribution of the bases would be lost. For example, the DNA might contain the following fractions of bases

$$p(C) = p(G) = 0.3$$
$$p(A) = p(T) = 0.2 \tag{2.24}$$

The entropy in this case would then be

$$S = -k(0.3 \ln 0.3 + 0.3 \ln 0.3 + 0.2 \ln 0.2 + 0.2 \ln 0.2)$$
$$= 1.36k \tag{2.25}$$

The entropy is reduced relative to the totally random DNA. The decrease in entropy as a system becomes more organized is sometimes defined as the increase in information I, that is,

$$I = 1.4k - 1.36k = 0.04k \tag{2.26}$$

The more formal definition of information will differ from the definition presented here $(I = \log_2 p)$, but the basic manipulations are essentially identical.

3. LEGENDRE TRANSFORMS

The first law of thermodynamics defines changes in internal energy in terms of changes in the entropy and volume,

$$dU = T\,dS - P\,dV \tag{2.27}$$

If the volume and entropy are held constant and no additional work terms are present, then the internal energy must also remain constant. This definition is particularly useful for a sealed container where V cannot change, as all changes in the internal energy must arise with the $T\,dS$ term.

It is often more convenient to work with open systems that are exposed to a constant pressure. For such systems it might be more appropriate to define an energy in which dP rather than dV was the relevant variable. The dP term could then be easily eliminated under constant pressure conditions. New energies with different differential variables can be generated using the Legendre transform described next. Because the product of any conjugate pair of variables has the units of energy, as for PV or TS, the transform must always deal with these conjugate pairs. In a Legendre transform, this conjugate pair is subtracted from the original energy to define a new energy. For example, the enthalpy, H, is formed by subtracting the product of $-P$ and V [Eq. (2.27)] from the internal energy U,

$$H = U - (-P)(V) = U + PV \tag{2.28}$$

The differential of the enthalpy is

$$\begin{aligned}
dH &= dU + P\,dV + V\,dP \\
&= T\,dS - P\,dV + P\,dV + V\,dP \\
&= T\,dS + V\,dP
\end{aligned} \tag{2.29}$$

This whole operation may appear rather arbitrary, but we have now defined an energy that is easily measured under constant pressure conditions to give

$$dH = T\,dS = dQ_P \qquad (2.30)$$

The enthalpy is then easily measured as the heat gained or lost at constant pressure.

The $T\,dS$ term that is present for both the internal energy and enthalpy expressions can be eliminated by maintaining constant entropy. However, it is far more convenient to define energies with a dT dependence, because a constant temperature situation is more experimentally accessible. The Legendre transform generates the Helmholtz free energy A and the Gibbs free energy G. The Helmholtz free energy is

$$A = U - (T)(S) = U - TS \qquad (2.31)$$

and yields

$$\begin{aligned} dA &= dU - T\,dS - S\,dT \\ &= T\,dS - P\,dV - T\,dS - S\,dT \\ &= -S\,dT - P\,dV \end{aligned} \qquad (2.32)$$

For the Gibbs free energy,

$$G = H - TS \qquad (2.33)$$

which yields

$$\begin{aligned} dH &= dH - T\,dS - S\,dT \\ &= -S\,dT + V\,dP \end{aligned} \qquad (2.34)$$

The Legendre transform has thus permitted us to define distinct energies for the system such that each has its own pair of independent variables, $U(S, V)$, $H(S, P)$, $A(T, V)$, and $G(T, P)$.

Although these four energies are standardized and commonly used, they are not the only transforms possible. Only PV work has been considered thus far, but a large number of additional work expressions are possible. In Chapter 1, the energy required to distort a surface was given as

$$\text{work} = \gamma\,dA \qquad (2.35)$$

where γ is the surface tension and dA is the change in the interfacial area. Energy can be used to expand the surface so that the first law becomes

$$dU = T\,dS - P\,dV + \gamma\,dA \qquad (2.36)$$

Note that the surface-energy term could be included with any of the energies defined thus far, for instance,

$$dG = -S\,dT + V\,dP + \gamma\,dA \tag{2.37}$$

Both T and P are intensive variables, whereas the surface area is an extensive variable. The Legendre transform permits us to generate a new free energy in terms of the intensive variables T, P, and γ:

$$G' = G - (\gamma)(A) = G - \gamma A \tag{2.38}$$

and

$$dG' = -S\,dT + V\,dP - A\,d\gamma \tag{2.39}$$

Note that γ is intensive, as it has units of energy per unit area.

Electrical energy is often expressed as the product of an electrical potential difference ψ multiplied by the amount of charge q that is transferred through this potential difference,

$$\text{energy} = \text{work} = q\psi \tag{2.40}$$

The potential has the units of joules per coulomb (energy per charge) and is the intensive variable in this case. To create a free energy with ψ as the independent variable, the free energy expression

$$dG = -S\,dT + V\,dP - \psi\,dq \tag{2.41}$$

is Legendre-transformed to

$$d\tilde{G} = -S\,dT + V\,dP + zF\,d\psi \tag{2.42}$$

where F represents the charge in coulombs for 1 mole of particles which each possess a net charge z. A variation of this particular energy will be used to establish a transmembrane potential in Chapter 4.

For all the energies discussed thus far, 1 mole of material has been assumed. However, the amount of material present also plays a role in the total energy of the system. For this reason, it is necessary to define an additional intensive–extensive pair of energy terms. The chemical potential of some substance i is defined as

$$\mu_i = (\partial G/\partial n_i)_{n_j,\,T,\,P} \tag{2.43}$$

and the energy is

$$\mu_i\,dn_i \tag{2.44}$$

The chemical potential appears to be just a free energy per mole and will reduce to this when only a single species is present. However, in general, the free energy will also depend on other substances, n_j, and the formal definition of Eq. (2.43) is essential.

4. MAXWELL'S RELATIONS

All the energies defined in Section 2.3 contain terms that are the product of an intensive with a conjugate extensive variable. The relations between such pairs are obvious. It is very advantageous to find relations between the variables of different conjugate pairs, and this can be done quite efficiently using Maxwell's relations.

The free energy involving only T and P is written as

$$dG = -S\,dT + V\,dP \tag{2.45}$$

Because dG is a state function, the free energy can also be expanded in terms of partial derivatives with respect to T and P,

$$dG = (\partial G/\partial T)_P\,dT + (\partial G/\partial P)_T\,dP \tag{2.46}$$

By comparing these two expressions for dG, we can establish the relations

$$(\partial G/\partial T)_P = -S \tag{2.47}$$

$$(\partial G/\partial P)_T = V \tag{2.48}$$

However,

$$\left[\frac{\partial}{\partial P}\left(\frac{\partial G}{\partial T}\right)_P\right]_T = \frac{\partial^2 G}{\partial T\,\partial P} = \left[\frac{\partial}{\partial T}\left(\frac{\partial G}{\partial P}\right)_T\right]_P \tag{2.49}$$

Substituting from Eqs. (2.47) and (2.48), we find

$$[\partial(-S)/\partial P]_T = (\partial V/\partial T)_P \tag{2.50}$$

as an example of a Maxwell relation. In general, we take the derivative of the nondifferential term in one conjugate expression with respect to the differential variable of the second conjugate pair.

Equation (2.50) can be used to illustrate the utility of the Maxwell relations. In order to determine some entropy change, an expression for the entropy is required. Equation (2.50) permits us to express the entropy in terms of more tractable variables. For an ideal gas,

$$\left(\frac{\partial V}{\partial T}\right)_P = \left[\frac{\partial}{\partial T}\left(\frac{RT}{P}\right)\right]_P = \frac{R}{P} \tag{2.51}$$

and the entropy is

$$-(\partial S/\partial P)_T = R/P \tag{2.52}$$

or

$$dS = -(R/P)\,dP \tag{2.53}$$

Integrating between pressures P_1 and P_2 then gives the entropy change for this process:

$$\Delta S = -R \ln(P_2/P_1) \tag{2.54}$$

The Maxwell relation thus permitted determination of the entropy in terms of other variables of the system.

A particularly interesting case arises when we develop Maxwell relations between the surface-energy term and the electrical potential term. The free energy would have the form

$$dG = -S \, dT + V \, dP - \gamma \, dA - \psi \, dq \tag{2.55}$$

This is now transformed to the new energy

$$dG'' = -S \, dT + V \, dP - \gamma \, dA + q \, d\psi \tag{2.56}$$

The last two terms in the expression now yield

$$[\partial(-\gamma)/\partial\psi]_A = (\partial q/\partial A)_\psi \tag{2.57}$$

The derivative on the right is the charge per unit area of surface, σ, so that

$$(\partial\gamma/\partial\psi)_A = -(\partial q/\partial A)_\psi = -\sigma \tag{2.58}$$

This is the Lippman equation. Maxwell's relations enable us to develop a relationship between the charge on the surface and the surface tension that is not immediately obvious.

5. THE GIBBS ADSORPTION ISOTHERM

Euler's theorem provides a second method of manipulating the variables of thermodynamic equations. The theorem is used to eliminate differentials for an equation in which all the differential quantities are extensive. For example, the internal energy U contains the four extensive variables S, V, n_i, and A in their differential forms,

$$dU = T \, dS - P \, dV + \gamma \, dA + \sum \mu_i \, dn_i \tag{2.59}$$

If each of these extensive variables were multiplied by a factor ξ, then the internal energy is also multiplied by ξ, because this factor can be brought in front of the entire expression on the left. Because ξ can be arbitrary, it can be made large enough to bring the differential terms to a macroscopic level. In other words, the extensive differential quantities can be replaced by the quantities themselves. This is Euler's theorem.

Equation (2.59) becomes

$$U = TS - PV + \gamma A + \sum \mu_i n_i \qquad (2.60)$$

This equation could now be differentiated again to produce the equation

$$dU = T\,dS + S\,dT - P\,dV - V\,dP + \gamma\,dA$$
$$+ A\,d\gamma + \sum \mu_i\,dn_i + \sum n_i\,d\mu_i \qquad (2.61)$$

This manipulation creates an interesting conflict, because both Eq. (2.59) and Eq. (2.61) are equal to dU. To resolve the conflict in a consistent fashion, the additional terms of Eq. (2.61) must sum to zero:

$$S\,dT - V\,dP + A\,d\gamma + \sum n_i\,d\mu_i = 0 \qquad (2.62)$$

Note that all differentials are intensive quantities. Because the extensive variables were selected as the independent variables to define the internal energy, Eq. (2.62) serves to place restrictions on the remaining dependent variables.

This version of the Gibbs–Duhem equation permits us to deduce one variable when we have information on the second. For example, at constant T, P, and γ for a two-component system, the Gibbs–Duhem equation is

$$n_1\,d\mu_1 = -n_2\,d\mu_2 \qquad (2.63)$$

or

$$d\mu_2 = -(n_1/n_2)\,d\mu_1 \qquad (2.64)$$

Thus, if the chemical potential μ_1 is known when n_1 and n_2 moles of the species are present, μ_2 can be determined using Eq. (2.64).

If the surface-energy term is also retained at constant T and P, the resultant equation for a two component system is

$$-A\,d\gamma = n_1\,d\mu_1 + n_2\,d\mu_2 \qquad (2.65)$$

Because the system can be defined for any region, this equation can be applied exclusively to the surface region so that n_2 and n_1 might now be the number of moles of solute and solvent, respectively. If surface concentration is now defined as moles per unit area, the equation becomes

$$-d\gamma = c_1\,d\mu_1 + c_2\,d\mu_2 \qquad (2.66)$$

where c_1 and c_2 will be the total concentrations of solvent and solute in the interfacial region.

For the two-component system, we often wish to know how the concentration of solute compares with this same concentration in bulk solution. If the surface concentration does differ from that of the bulk concentration, we can define a concentration or surface excess. In order to do this, the concentration of solvent should be the same for both the surface and bulk phases. The boundary that separates bulk phase from interfacial phase is chosen accordingly. For example, if the bulk solution contained 0.1 moles of solute and 1 mole of solvent, a surface region containing 1 mole of solvent for unit area of surface would be selected. If the solute concentration for this surface region was 0.3 moles, then the surface excess would be 0.2 moles per unit area. Since the surface phase has the same number of moles of solvent as the bulk phase, there is no change in μ_1, that is, solvent becomes homogeneous with this choice of phase boundary, and

$$d\mu_1 = 0 \tag{2.67}$$

If the excess surface concentration is defined as Γ_2, Equation (2.66) becomes

$$-d\gamma = \Gamma_2 \, d\mu_2 \tag{2.68}$$

Measurement of the surface tension for a number of different bulk solute concentrations will give the excess concentrations in the interfacial regions. Note that "excess" may be positive or negative. Equations (2.66) and (2.68) describe two types of concentration definition for the surface region. Other definitions are possible.

In Chapter 4, the chemical potential of the solute will be related to the solute concentration via the relation

$$-d\mu = RT \, d(\ln c) \qquad \text{or} \qquad RT \, d(\ln a) \tag{2.69}$$

If this expression is substituted into Eq. (2.68), the Gibbs adsorption isotherm results:

$$-d\gamma = \Gamma_2 RT \, d(\ln c) \tag{2.70}$$

6. BOLTZMANN STATISTICS

The entropy for a set of molecules in a sealed container is

$$S = k \ln \Omega \tag{2.71}$$

as derived in Section 2.2. This equation may be written in the alternative form

$$\Omega = \exp(S/k) = \exp(TS/kT) \tag{2.72}$$

where TS is the reversible energy associated with the molecules having an entropy S. This form suggests that exponentials of the form

$$\exp(-\varepsilon/kT) \tag{2.73}$$

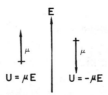

Fig. 2.3 Two extreme dipole orientations in an electric field. The aligned dipole is the lower energy configuration.

(where ε is some general energy) can provide a measure of the total number of configurations generated by that energy. For example, if ε is a free energy,

$$\varepsilon = G = H - TS \tag{2.74}$$

then the exponential factor is

$$\exp[-H/kT]\,\exp[+TS/kT] \tag{2.75}$$

The second factor describes the number of possible arrangements (that is, Ω), while the initial term reflects the fact that the energy of a state relative to the thermal energy, kT, of that state mediates the probability of observing that state.

This Boltzmann factor permits us to define relative populations for each state and the thermodynamic properties of the system. To illustrate, consider the analysis for the simple dipole model shown in Fig. 2.3. In Chapter 3, the energy for a dipole with dipole moment μ is determined to be

$$U = -\boldsymbol{\mu} \cdot \mathbf{E} = -E\mu \cos \theta \tag{2.76}$$

where the angle θ can assume any value from $0°$ to $180°$. However, consider the situation for which only the two dipole orientations shown in Fig. 2.3 are possible. The two energies are then

$$U_1 = -\mu E$$

$$U_2 = +\mu E \tag{2.77}$$

where the negative energy state is the lower energy state of the two. The probabilities of finding the dipole in each of these two states will be proportional to the Boltzmann factors for these states,

$$p_1 \propto \exp[-(-\mu E)/kT) = \exp(\mu E/kT) \tag{2.78}$$

$$p_2 \propto \exp(-\mu E/kT) \tag{2.79}$$

To determine the actual probability that the dipole is in state 1, Eq. (2.78) must be divided by the sum of both probabilities,

$$p_1 = \frac{\exp(\mu E/kT)}{\exp(\mu E/kT) + \exp(-\mu E/kT)} \tag{2.80}$$

Also,

$$p_2 = \frac{\exp(-\mu E/kT)}{\exp(\mu E/kT) + \exp(-\mu E/kT)} \tag{2.81}$$

The common factor in the denominators of Eqs. (2.80) and (2.81) ensures that the sum of all probabilities for the system is 1:

$$p_1 + p_2 = 1 \tag{2.82}$$

This is called the partition function or sum over states for this particular system. For systems with a larger number of states with different energies, the sum over states will be extended to sum the Boltzmann factors for all these states. The symbol Q will be used for such sums involving only the molecular energy.

To illustrate this approach, consider the special case where $|\mu E| = kT$. The two probabilities are then

$$p_1 = \exp(+1)/Q = 2.72/Q \tag{2.83}$$

and

$$p_2 = \exp(-1)/Q = 0.37/Q \tag{2.84}$$

The probability—that is, the fraction of all the dipoles in configuration 1—is

$$p_1 = 2.72/(2.72 + 0.37) = 0.88 \tag{2.85}$$

and

$$p_2 = 0.37/3.09 = 0.12 \tag{2.86}$$

A solid majority of the dipoles will assume the lower energy orientation.

The probabilities for each of the energetic states permit us to determine average values for a variety of parameters. The average value for some quantity X that has values X_i for each of the energy states i will be

$$\langle X \rangle = \sum p_i X_i \tag{2.87}$$

For the two-state model, the average energy will be

$$\langle U \rangle = [(-\mu E) \exp(\mu E/kT)]/Q + (\mu E) \exp(-\mu E/kT)$$

$$= (\mu E) \frac{[\exp(-\mu E/kT) - \exp(\mu E/kT)]}{[\exp(-\mu E/kT) + \exp(\mu E/kT)]} \tag{2.88}$$

Both numerator and denominator are reasonably similar, as both are sums of the same exponential functions. This suggests it may be possible to express the numerator in terms of the denominator. To illustrate the approach more clearly, let

$$\beta = 1/kT \qquad (2.89)$$

so that the denominator becomes

$$\exp(-\beta\mu E) + \exp(\beta\mu E) \qquad (2.90)$$

If this sum over states is now differentiated with respect to β while holding μ and E constant in each term, and is then multiplied by -1, we find

$$\frac{-\partial}{\partial\beta}\ [\exp(-\beta\mu E) + \exp(\mu\beta E)] = (\mu E)\exp(-\beta\mu E) + (-\mu E)\exp(\beta\mu E) \qquad (2.91)$$

The derivative of the sum over states with respect to β produces the numerator for the energy expression. Equation (2.88) can now be written as

$$\langle U \rangle = \left[\frac{-\partial Q}{\partial\beta}\right]\left(\frac{1}{Q}\right) \qquad (2.92)$$

$$= -\partial(\ln Q)/\partial\beta \qquad (2.93)$$

This equation can be checked by performing the operations to regenerate Eq. (2.88). Although this manipulation appears to give no new information for this simple two-state case, it becomes extremely important when a large number of distinct states are present. If the partition function can be summed, Eq. (2.93) then permits a rapid calculation of the average energy for all these states.

The same partition function could also be used to find the average dipole moment for the system. From Eq. (2.87), this is

$$\langle \mu \rangle = \frac{\mu\exp(-\beta\mu E) + (-\mu)\exp(\beta\mu E)}{\exp(-\beta\mu E) + \exp(\beta\mu E)} \qquad (2.94)$$

The reader should confirm that this can be reduced to the following differential expression:

$$\langle \mu \rangle = -\partial(\ln Q)/\partial(\beta E) \qquad (2.95)$$

7. THE LANGEVIN EQUATION

The average energy and dipole moment of dipoles in an electric field were determined for the special case in which only two configurations were possible. However, the dipole will generally be free to assume any orientation

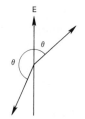

Fig. 2.4 Two dipole orientations from the continuum of orientations for the dipoles $0 \leq \theta \leq 180°$.

in the field, as shown in Fig. 2.4. The energy associated with the angle θ is

$$U(\theta) = -\mu E \cos \theta \qquad (2.96)$$

and the Boltzmann factor for this energy is

$$\exp(+\beta \mu E \cos \theta) \qquad (2.97)$$

where $0° < \theta < 180°$. To determine the partition function Q for this case, the Boltzmann factors for each value of θ must be summed. Since the angles vary continuously, an integral must be used to determine the resultant sum over states. Once Q is known, it can be used to determine both the energy and the average dipole moment for the dipoles.

In order to sum the contributions properly in this polar coordinate system, we must integrate with respect to $\sin \theta \, d\theta$ rather than $d\theta$. The fact that this element represents a "distance" is illustrated in Fig. 2.5. Because the energy

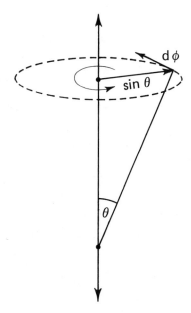

Fig. 2.5 Source of the $\sin \theta$ factor in the Langevin integral. The $\sin \theta$ is required to convert $d\phi$ to a length in spherical coordinates.

will not vary with the equatorial angle ϕ, this factor need not be included, as it will be common to both numerator and denominator.

The integral for Q is

$$Q = \int_0^{180} \exp(\beta\mu E \cos\theta) \sin\theta \, d\theta \tag{2.98}$$

This integral can be simplified by defining

$$y = \beta\mu E \cos\theta \tag{2.99}$$

and

$$dy = -\beta\mu E \sin\theta \, d\theta \tag{2.100}$$

The limits of the integral must also be changed. When $\theta = 0$, $y = +\beta\mu E$. When $\theta = 180°$, $y = -\beta\mu E$. The integral is then

$$Q = (\beta\mu E)^{-1} \int_{(+\beta\mu E)}^{(-\beta\mu E)} [\exp(y) \, dy](-1) \tag{2.101}$$

$$= (\beta\mu E)^{-1}(-\exp y)\Big|_{(\beta\mu E)}^{(-\beta\mu E)} \tag{2.102}$$

and so

$$Q = (\beta\mu E)^{-1}[\exp(\beta\mu E) - \exp(-\beta\mu E)] \tag{2.103}$$

$$\ln[Q] = \ln[\exp(\beta\mu E) - \exp(-\beta\mu E)] - \ln(\beta\mu E) \tag{2.104}$$

The average energy is

$$\partial \ln(Q)/\partial\beta = \frac{E[\exp(\beta\mu E) + \exp(-\beta\mu E)]}{[\exp(\beta\mu E) - \exp(-\beta\mu E)]} - \frac{\mu E}{\beta\mu E} \tag{2.105}$$

$$= \mu E \coth(\beta\mu E) - \frac{1}{\beta} \tag{2.106}$$

or

$$U = \langle\mu\rangle E = \mu E\left[\coth(\beta\mu E) - \frac{1}{\beta\mu E}\right] \tag{2.107}$$

since the electric field E is constant for all dipoles. The average dipole moment can be determined by dividing the energy expression by E,

$$\langle\mu\rangle = \mu[\coth(\beta\mu E) - (1/\beta\mu E)] \tag{2.108}$$

This is the Langevin equation for an average dipole moment of a dipole in an electric field E.

The Langevin equation simplifies when the dipole is placed in a small electric field so that $\mu E < kT$. For this limiting case, the average dipole moment reduces to

$$\langle \mu \rangle = \mu^2 E/3kT \tag{2.109}$$

The derivations of this limiting case and the limiting case for the two-state model [Eq. (2.94)] are left as exercises for the reader. The two-state approximation yields a limiting value of

$$\langle \mu \rangle = \mu^2 E/kT \tag{2.110}$$

which is a factor of three larger than the Langevin case, since the two-state model tends to overemphasize the lower energy state.

8. THE GRAND CANONICAL PARTITION FUNCTION AND MEMBRANE BINDING

The phospholipids of a bilayer membrane have their polar region at the membrane–solution interface. In this configuration, the molecule may acquire different energies U_i. If we wished to characterize the thermodynamic properties of this molecule, our starting point would be the sum over states, that is, the sum of all Boltzmann factors containing the different energies possible for the molecule. This partition function is then

$$q = \sum \exp(-U_i/kT) \tag{2.111}$$

If the membrane region of interest contained N such membrane molecules and these molecules all had energy levels that were not perturbed by the neighboring molecules, the partition function for these N molecules would be

$$Q = q^N \tag{2.112}$$

To illustrate this point, consider the case for two such molecules with two possible energies per molecule (U_1 and U_2). There are four distinct arrangements:

Molecule 1	Molecule 2	
$\exp(-U_1/kT)$	$\exp(-U_1/kT)$	
$\exp(-U_1/kT)$	$\exp(-U_2/kT)$	(2.113)
$\exp(-U_2/kT)$	$\exp(-U_1/kT)$	
$\exp(-U_2/kT)$	$\exp(-U_2/kT)$	

The total energy for the first pair above is $U_1 + U_1$, and its Boltzmann factor is

$$\exp(-2U_1/kT) = \exp(-U_1/kT)\exp(-U_1/kT) \qquad (2.114)$$

The total Boltzmann factor for each biomolecular configuration is formed as the product of factors for the individual molecules. The total sum over states is

$$\exp(-2U_1/kT) + 2\exp[-(U_1 + U_2)/kT] + \exp(-2U_2/kT) \qquad (2.115)$$

which is equivalent to

$$[\exp(-U_1/kT) + \exp(-U_2/kT)]^2 = q^2 \qquad (2.116)$$

When each of the molecules has an independent set of energy levels, we can form the total partition function by raising the partition function for a single molecule to the appropriate power, that is, the number of molecules. This suggests we might develop more complicated partition functions for the individual molecules; this is generally less complicated than developing the partition function for the entire surface directly.

Consider the single membrane molecule, shown in Fig. 2.6, facing a solution containing a concentration c of the cation C^+, which is the only species in solution available to bind to the molecule. Only two "states" are now possible. The unbound molecule with all its specific energy states is characterized by the partition function q. The second state with the bound ion has the energy states of the molecule, so it also requires the partition function q. In addition, it must include a Boltzmann factor relating to the energy of binding of the cation,

$$\exp(-E_b/kT) = q_b \qquad (2.117)$$

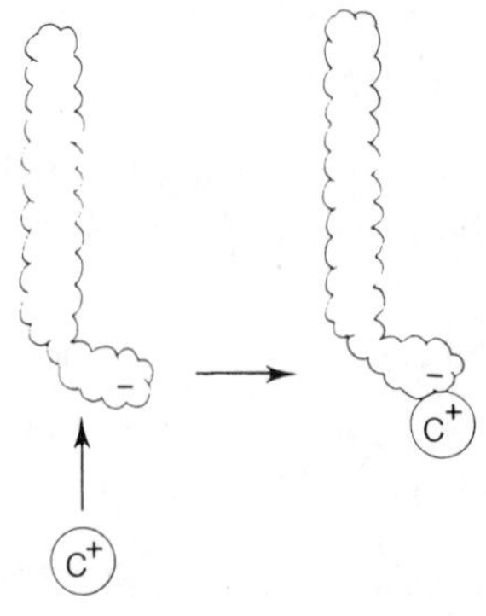

Fig. 2.6 Bound and unbound configurations for a single membrane molecule.

and an energy that characterizes the ability of the cation to move from the solution to the molecule. This energy is the chemical potential difference μ between the solution and the molecule. The Boltzmann factor is

$$\lambda = \exp(-\mu/kT) \qquad (2.118)$$

When the chemical potential of the cation on the molecule is lower than the chemical potential of the cation in solution ($\mu < 0$), the Boltzmann factor for the bound species will be enhanced relative to that of the unbound molecule.

The total partition function for the molecule-cation system is

$$\xi = q + qq_b\lambda = q(1 + q_b\lambda) \qquad (2.119)$$

where ξ is the partition function for a molecule with one binding site. The grand canonical partition function for all N sites is then

$$\Xi = \xi^N \qquad (2.120)$$

To determine the fraction of all sites that have bound cation, we consider the two terms of Eq. (2.117). The sum includes all possible states of the system, but the second term characterizes those states with bound cation. The fraction of sites with bound cation is then

$$\begin{aligned} f_b &= qq_b\lambda/(q + qq_b\lambda) \\ &= q_b\lambda/(1 + q_b\lambda) \end{aligned} \qquad (2.121)$$

This result could be derived using the general formula

$$b_b = \frac{\partial \ln \xi}{\partial \lambda} \qquad (2.122)$$

The derivative with respect to λ effectively places all the cation-bound terms in the numerator.

The partition function can be extended to include binding for any number of different ions; it can also be extended to molecules containing more than one potential binding site. In all cases, the approach is the same. Partition functions are formed for each of the distinct states. These partition functions are then summed.

Consider the situation described in Fig. 2.7. The molecule has two binding sites that can bind either cation 1 or cation 2. The distinct possibilities are tabulated in the figure. Given that the sites are assumed to be equivalent, the partition function is

$$\xi = q(1 + 2\lambda_1 q_1 + 2\lambda_2 q_2 + \lambda_1^2 q_1^2 + \lambda_2^2 q_2^2 + 2\lambda_1\lambda_2 q_1 q_2) \qquad (2.123)$$

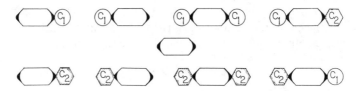

Fig. 2.7 Different binding configurations possible when two different cations C_1 and C_2 compete for two identical binding sites.

The fraction of bound cation 1 is

$$\lambda_1 \frac{\partial \ln \xi}{\partial \lambda_1} = q(2\lambda_1 q_1 + 2\lambda_1^2 q_1^2 + 2\lambda_1 \lambda_2 q_1 q_2)/\xi \qquad (2.124)$$

The derivative picks those terms containing λ_1. It also introduces a factor of 2 for the term in which two "1" cations are bound. The bound fractions obtained using the partition function approach are identical to those that would be determined from chemical equilibrium calculations. If a single cation can bind to the molecule, the equilibrium constant is

$$K = [CS]/[S^-][C^+] \qquad (2.125)$$

where $[S] = [CS] + [S^-]$ is the total concentration of binding sites and $[C^+]$ is the cation concentration in solution. Because

$$[CS] = K[S^-][C^+] \qquad (2.126)$$

we have in turn

$$[S] = [S^-] + K[S^-][C^+] \qquad (2.127)$$

The fraction bound is then determined by the concentration $[CS]$ divided by $[S]$, the total site concentration:

$$\begin{aligned}
f_b &= [CS]/[S] \\
&= K[S^-][C^+]/([S^-] + K[S^-][C^+]) \\
&= K[C^+]/(1 + K[C^+]) \qquad (2.128)
\end{aligned}$$

The concentration $[C^+]$ determines the driving force for moving cations from solution to the binding sites and is therefore compatible with λ. The equilibrium constant K reflects the free energy difference between the cation in solution and bound to the site and is related to the partition function q_b. Thus, the expressions derived using statistical thermodynamics and thermodynamics are essentially identical.

PROBLEMS

1. The internal energy of a system is

$$dU = T \, dS - P \, dV - \gamma \, dA$$

Develop an expression for a new thermodynamic energy, $J = J(T, V, \gamma)$.

2. The energy for an ionic solution is

$$dG = -S \, dT + V \, dP - Q \, d\psi$$

Develop differential and integral expressions for the entropy change produced by a change in the potential ψ.

3. The Curie–Weiss law for ferromagnetic materials relates the magnetic field H and the temperature T to an induced magnetization M,

$$M = CH/(T - T_c)$$

where C and T_c are constants. The work required for a change of magnetization dM is $-H \, dM$. (a) Write an expression for the Helmholtz free energy dA. (b) Determine the work for magnetization when the magnetic field is changed from H_1 to H_2 using the Curie–Weiss Law. (c) Use Maxwell's relations to determine the entropy change when the magnetic field is changed from H_1 to H_2.

4. Charged sites on a membrane surface are arranged in pairs, so that solution ions can bind on either site. In addition, one univalent ion can bind to each site. In this case, there is an additional repulsive energy E_r. (a) Write the grand canonical partition function for this system when there are a total of N pairs of sites. (b) Determine the fraction of sites with bound ions.

5. For the membrane binding sites of Problem 4, determine the grand partition function when both a univalent and a divalent species are present in solution. Determine the fraction of sites with bound divalent cation.

6. Show that the Langevin function reduces to the form

$$\langle \mu \rangle = \mu(\mu E/3kT)$$

for small electric fields.

Chapter 3

Electrical Interactions

1. ELECTRICAL UNITS

The definition of electrical charge differs for the cgs and MKS systems, because different fundamental equations are used to define charge for the two systems. The cgs system defines charge using Coulomb's law,

$$F = -q_1 q_2 / r^2 \tag{3.1}$$

In cgs units, r has the units of centimeters and force has the units of dynes, and the units of q_1 and q_2 are selected so that a separation of 1 cm would produce an attraction or repulsion of 1 dyn. If the two charges have equal magnitude the unit of charge that satisfies this condition is 1 esu (electrostatic unit).

An accurate determination of charge using Coulomb's law as the operational definition requires an accurate measurement of both distance and force. For this reason, an alternative charge measurement is preferable. Charge flowing in a circuit produces a magnetic field, which then exerts a measurable force. The force between this charge flow (current) and a second, known, current flow is used to determine the charge. The charge in MKS units defined in this manner is the coulomb. An additional constant is required in Coulomb's law so that charge can be expressed in coulombs (C). Coulomb's law is now

$$F = +(1/4\pi\varepsilon_0)(q_1 q_2 / r^2) \tag{3.2}$$

where the constant ε_0 is called the permittivity of free space 8.85 × 10^{-12} C^2/J · m) or 8.85 × 10^{-12} F/m). The two statements of Coulomb's law can be compared by determining the esu equivalent value for the charge

44

on an electron (1.6×10^{-19} C). Using Eq. (3.2), two such charges will produce a force of 2.3×10^{-28} N (newtons) at a distance of 1 m. In cgs units, this is equivalent to a force of 2.3×10^{-23} dyn at a distance of 100 cm. The esu charge on the electrons is then

$$(esu)^2 = [force(dyn)][distance(cm)]^2$$
$$= 2.3 \times 10^{-23} \times 10^4 = 23 \times 10^{-20} \ esu^2 \quad (3.3)$$

The electronic charge is cgs units is then 4.8×10^{-10} esu.

The MKS units of charge are the central units in this text. However, because much of the literature uses cgs units, parallel definitions are given. Generally, the only major difference between equations developed in the different unit systems will involve the addition or deletion of the factor $4\pi\varepsilon_0$.

2. ALTERNATIVE FORMS OF COULOMB'S LAW

Coulomb's law as defined by Eq. (3.1) or (3.2) involved the interaction between two charged particles. In many cases, it is convenient to express this force in terms of one of the charged particles. In such cases, we define an electric field E as

$$E = F/q \quad (3.4)$$

The electric field is then the force F that a given charge q would exert on a unit test charge. For any arbitrary charge, the force would be the product of the electric field and this charge,

$$F = Eq_i \quad (3.5)$$

The electric field has units of newtons per coulomb (N/C).

If two particles experience an attractive force, they will be drawn together and work will be done. It is often convenient to determine the work as the energy produced when a test charge moves due to another charge (or charges). A potential will be the energy per unit charge,

$$V = (W/q) = (J/C) \quad (3.6)$$

The electric field can be expressed in an alternative form, as 1 joule = 1 newton-meter.

$$E = (N/C)(m/m) = (N \cdot m/C)/m = V/m \quad (3.7)$$

For a single particle of charge q, the electric field is defined using Coulomb's law

$$E = F/q_t = \frac{(1/4\pi\varepsilon_0)(qq_t/r^2)}{q_t} = (1/4\pi\varepsilon_0)(q/r^2) \quad (3.8)$$

This equation can be written as

$$E(4\pi r^2) = q/\varepsilon_0 \qquad (3.9)$$

This observed field through the surface $4\pi r^2$ is proportional to the total charge q contained within that surface. Using vector notation, this result can be generalized to a surface of arbitrary shape

$$\int \mathbf{E} \cdot d\mathbf{S} = q/\varepsilon_0 = \sum q_i/\varepsilon_0 \qquad (3.10)$$

where $\int$ indicates an integral over the entire surface. All the charges q_i within the surface-enclosed region will contribute to the electric field. This form of Coulomb's law is called Gauss's law.

To illustrate Gauss's law, consider a cylinder in an electric field along the x axis, as shown in Fig. 3.1. There are no charges within the cylinder. The electric field will be parallel only to the two vectors perpendicular to the two cylinder faces. These are the only portions of the surface that can contribute to the surface integral. The surface integral is then

$$\mathbf{E} \cdot \mathbf{A}_1 + \mathbf{E} \cdot \mathbf{A}_2 = -EA_1 + EA_2 = 0 \qquad (3.11)$$

because the normals to the cylinder faces face in opposite directions (Fig. 3.1). In other words, the electric field is not altered as it passes through the cylinder because there is no charge inside.

Now, consider a cylinder of length Δx and assume this cylinder is filled with charge, so that the total charge within the cylinder must be expressed as a density ρ of charge (charge per unit volume),

$$q_t = \rho(\Delta x A) \qquad (3.12)$$

If the electric field passes through this cylinder, this additional charge should add to the electric field, so that the electric field leaving the right face, E_2,

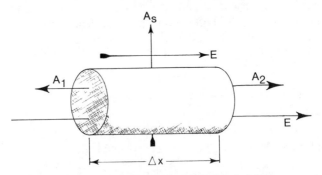

Fig. 3.1 Gauss's law for a cylindrical region. The electric field components for the two faces are equal and opposite. Components from the cylindrical surface are perpendicular to the field.

is now larger than E, the electric field entering the left face. This increase can be expressed as a Taylor series,

$$E_2 = E + [dE/dx]\Delta x \tag{3.13}$$

and the net difference in fields through the opposite cylinder faces is

$$E_2 A_2 - EA_1 = ([dE/dx]\Delta x)A \tag{3.14}$$

From Gauss's law, this net field must be equal to the charge in the cylinder, so

$$([dE/dx]\Delta x)A = \rho \Delta x A/\varepsilon_0 \tag{3.15}$$

Cancelling common terms yields

$$dE/dx = \rho/\varepsilon_0 \tag{3.16}$$

This differential form of Gauss's and Coulomb's law is called the Poisson equation.

Poisson's equation can be derived more elegantly in three dimensions using the divergence theorem,

$$\int \mathbf{E} \cdot d\mathbf{S} = \int \mathbf{\nabla} \cdot \mathbf{E} \, dV \tag{3.17}$$

which converts the surface integral to a volume integral. The total charge within the volume is

$$\int \rho \, dV \tag{3.18}$$

and Gauss's law becomes

$$\int \mathbf{E} \cdot d\mathbf{S} = \int \mathbf{\nabla} \cdot \mathbf{E} \, dV = \int \rho \, dV/\varepsilon_0 \tag{3.19}$$

Since the two volume integrals are equal, their arguments must be equal, so that

$$\mathbf{\nabla} \cdot \mathbf{E} = \rho/\varepsilon_0 \tag{3.20}$$

This is the three-dimensional version of Poisson's equation.

When the electrical potential varies with distance, the electric field associated with this potential is defined via a derivative,

$$E = -dV/dx \tag{3.21}$$

for one dimension or

$$\mathbf{E} = -\mathbf{\nabla} V \tag{3.22}$$

for three dimensions.

Poisson's equation can now be expressed in terms of potential

$$dE/dx = -d^2V/dx^2 = \rho/\varepsilon_o \tag{3.23}$$

or

$$\nabla^2 V = -\rho/\varepsilon_o \tag{3.24}$$

It is often convenient to express the coulomb energy in kJ/mole to facilitate comparison with other energies. The Coulomb expression can be written as

$$U = 1391 n_1 n_2/\varepsilon_r r \text{ kJ/mole} \tag{3.25}$$

where n_1 and n_2 are the numbers of charges on each of the interacting species and r is the distance between them in angstroms; ε_r is a relative dielectric constant, with $\varepsilon_r = 1$ for vacuum, and $\varepsilon_r = 80$ for water. Thus, a hydrated K^+ ion at a distance of 5 Å from a membrane surface site would generate an interaction energy of 3.48 kJ/mole. For the unhydrated ion at its distance of closest approach (1.5 Å, $\varepsilon_r = 1$), this energy will increase to 927 kJ/mole.

3. ELECTRIC DIPOLES

An electric dipole consists of two equal and opposite charges separated by some distance d. Obviously such charges would attract each other, so they must be prevented from doing so by some intrinsic rigidity for the system. For example, the charges could be localized in different regions of a molecule. The potential produced by a dipole will be different from that of a single charge, since the effects of the two oppositely charged regions must be considered. To illustrate, consider the potential that might be expected if a test charge is located at some distance r from the center of the dipole and perpendicular to the dipole axis, as shown in Fig. 3.2. To determine the net

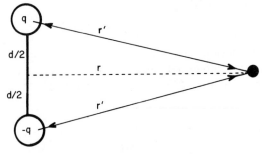

Fig. 3.2 Net dipole potential on a test charge equidistant (r') from $+q$ and $-q$, where r is the dipole–test charge separation.

potential experienced by the test charge, the potentials produced by each of the charges can be summed. The distance to each of the dipole charges will be

$$r' = [r^2 + (d/2)^2]^{1/2} \tag{3.26}$$

and the net potential experienced by the test charge is

$$V_{net} = [1/4\pi\varepsilon_0][(|q|/r') + (-|q|/r')] = 0 \tag{3.27}$$

because the charges have opposite sign. The effects of the two charges cancel everywhere (all r) on this perpendicular axis.

If the test charge is located along the dipole axis at some distance r from the center of the dipole as shown in Fig. 3.3, the positive dipole charge will always be closer to the test charge and a net repulsion is expected. However,

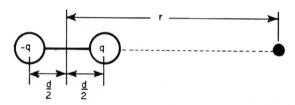

Fig. 3.3 Net dipole potential for a test charge with separation r along the dipole axis coordinate, when d is the distance between dipole charges.

for large r relative to d, the small apparent separation of the charges will produce very little change in the two potentials. The dipole potential will be short range. As shown in Fig. 3.3, the positive charge will be $r - (d/2)$ from the test charge, while the negative charge will be $r + (d/2)$ away. The net potential is

$$V = [1/4\pi\varepsilon_0\varepsilon_r]\left[\frac{1q}{r - (d/2)} + \frac{-1q}{r + (d/2)}\right] \tag{3.28}$$

Now r is factored from each term in the denominator

$$V = \frac{q}{4\pi\varepsilon_0\varepsilon_r r}\left[\frac{1}{1 - (d/2r)} - \frac{1}{1 + (d/2r)}\right] \tag{3.29}$$

The expression is expanded by noting that

$$(1 - x)^{-1} = 1 + x + x^2 + \cdots \tag{3.30}$$

$$(1 + x)^{-1} = 1 - x + x^2 - x^3 + \cdots \tag{3.31}$$

so that

$$V = [q/4\pi\varepsilon_o\varepsilon_r r][1 + (d/2r) + (d/2r)^2 + \cdots - 1 + (d/2r) - (d/2r)^2 + \cdots]$$
$$\text{(3.32)}$$

$$= [q/4\pi\varepsilon_o\varepsilon_r r](2d/2r) = \frac{qd}{4\pi\varepsilon_o\varepsilon_r r^2} \tag{3.33}$$

The numerator, $\mu = qd$, is the dipole moment; it is the product of the absolute value of the separated charges and the distance of separation. In MKS units, the dipole moment has the units of $C \cdot m$. In cgs units, its units are $esu \cdot cm$. For molecular systems, the dipole moments involve single electronic charges separated by angstroms. For a 1-Å separation,

$$\mu = (4.8 \times 10^{-10})(1 \times 10^{-8}\ cm) = 4.8 \times 10^{-18}\ esu\ cm \tag{3.34}$$

Since the factor 10^{-18} appears consistently with cgs units, it is defined in 1 debye, such that

$$10^{-18}\ esu \cdot cm = 1\ debye \tag{3.35}$$

Although the dipole potential was derived from the Coulomb potential with its inverse r dependence, the final potential expression depends inversely on r^2. The interaction decreases more rapidly with r, because at large distances the positive and negative charge appear as a net "neutral."

If the test charge is now placed on the dipole axis so that it faces the negative pole, the equation is identical to Eq. (3.33) with the sign reversed,

$$V = -[q/4\pi\varepsilon_o\varepsilon_r r](d/r) = \frac{-\mu}{4\pi\varepsilon_o\varepsilon_r r^2} \tag{3.36}$$

The three limiting cases suggest the general form for the dipole potential when the test charge is found along an axis at an angle θ to the dipole axis, as shown in Fig. 3.4. The general expression is

$$V = -\mu \cos \theta/4\pi\varepsilon_o\varepsilon_r r^2 \tag{3.37}$$

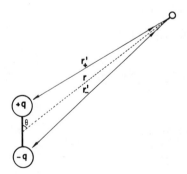

Fig. 3.4 Net dipole potential for a test charge aligned at an angle θ with respect to the dipole axis.

This general expression can be developed using the law of cosines,

$$(r')^2 = r^2 + (d/2)^2 - 2r(d/2) \cos \theta \qquad (3.38)$$

to determine the distances to each test charge. This is left as an exercise.

The electric field of the dipole is determined by taking the derivative of the potential. However, there are now two independent variables in the expression for the potential, r and θ. Each of these variables can be used to define force components. The electric field component in the radial direction is

$$E_r = -\partial V/\partial r = 2\mu \cos \theta/4\pi\varepsilon_o\varepsilon_r r^3 \qquad (3.39)$$

For $\theta = 90°$, there is no electric field.

The electric field component for θ is

$$E_\theta = -r^{-1}\partial V/\partial\theta = \mu \sin \theta/4\pi\varepsilon_o\varepsilon_r r^3 \qquad (3.40)$$

because the r^{-1} term is an intrinsic part of the gradient in spherical coordinates.

The equation for the dipole potential [Eq. (3.37)] contains a number of constant terms, so it is again convenient to reexpress the equation in kJ/mole. The energy of interaction between an ion with n electronic charges and the dipole is

$$U = V(ne) \qquad (3.41)$$

and this is written in the form

$$U = 6655.2nd \cos(\theta)/\varepsilon_r r^2 \text{ kJ/mole} \qquad (3.42)$$

where d is the separation between dipole charges in angstroms, ε_r is the relative dielectric constant of the medium, and r is the distance to the charge in angstroms.

For a dipole with charge separation of 5 Å, $\theta = 0°$, and a distance of $r = 10$ Å to a univalent ($n = 1$) ion in water, we find an energy

$$U = (6655.2)(5)/[80(10)^2] = 1.73 \text{ kJ/mole} \qquad (3.43)$$

The r^2 dependence evidently plays a major role in reducing this energy relative to pure coulombic interactions. The 1.7 kJ/mole value compares with a thermal energy of

$$RT = 2.48 \text{ kJ/mole}$$

at 25°C. For the 10 Å separation, the thermal effects will dominate the dipole–charge interaction.

4. DIPOLE ROTATION IN ELECTRIC FIELDS

A dipole will rotate when placed in a homogeneous electric field. The negative component of the dipole will move in the negative field direction, while the positive component will move in the opposite direction, as shown in Fig. 3.5. These interactions will produce a torque on the dipole to align it in its most stable configuration. For rotation about the dipole center, the total torque acting on the dipole perpendicular to the field is

$$T = qE(d/2) - [-qE(d/2)] = (qd)E = \mu E \qquad (3.44)$$

for the case where the angle between the electric field and the dipole is 90°. For an angle θ from the electric field direction, the component which experiences the torque is

$$(d/2) \sin \theta \qquad (3.45)$$

and the torque is

$$T = \mu E \sin \theta \qquad (3.46)$$

The energy required to orient the dipole in the field at an angle θ can be determined by introducing the two charges in an overlapping (neutralized) configuration and then separating them in the field. Consider the limiting case of $\theta = 90°$. As the charges separate, they both move perpendicular to the field and no work is done.

If the charges are now separated parallel to the field so that the positive charge moves in the field direction, each charge will be moving toward its most stable configuration and energy will be released. The total work is the product of the charge, the field, and the distance travelled. Since each charge moves a distance $d/2$, the energy is

$$U = -qE(d/2) + qE(-d/2) = -q\,dE = -\mu E \qquad (3.47)$$

If the charges are moved in the opposite direction, they require work against the field,

$$U = +\mu E \qquad (3.48)$$

Fig. 3.5 Torques experienced by a dipole in a constant electric field. The dipole axis is perpendicular to the electric field.

For an arbitrary angle θ between the field and the dipole, the energy is

$$U = -\mu E \cos \theta = -\boldsymbol{\mu} \cdot \mathbf{E} \qquad (3.49)$$

To determine the energy produced by rotation of the dipole in an electric field, the fields generated across membrane systems are considered. Such membranes can sustain potentials of 0.150 V and, for a 60-Å membrane, the homogeneous field is

$$E = -\psi/L = 0.15/(0.6 \times 10^{-8} \text{ m}) = 2.5 \times 10^7 \text{ V/m} \qquad (3.50)$$

For a dipole with a charge separation of 5 Å, the dipole moment is

$$\mu = qd = (1.6 \times 10^{-19} \text{ C})(5 \times 10^{-10} \text{ m}) = 8 \times 10^{-29} \text{ C} \cdot \text{m} \qquad (3.51)$$

In cgs units, this dipole moment is

$$\mu = qd = (4.8 \times 10^{-10} \text{ esu})(5 \times 10^{-8} \text{ cm}) = 24 \text{ debye} \qquad (3.52)$$

The energy required to rotate the dipole from $\theta = 0°$ to $\theta = 180°$ is

$$U = \mu E - (-\mu E) = 2\mu E$$
$$= (2)(8 \times 10^{-29})(2.5 \times 10^7) = 4.0 \times 10^{-21} \text{ J/molecule} \qquad (3.53)$$

For 1 mole of particles,

$$U = 2.40 \text{ kJ/mole} \qquad (3.54)$$

which is comparable to the thermal energy of 2.5 kJ/mole. Thus, such fields can influence the orientation of the dipole at room temperature.

5. DIPOLE–DIPOLE INTERACTIONS

In addition to interactions with charges and electric fields, dipoles can also interact with other dipoles. Such interactions can be important, because they may serve to stabilize membrane arrays.

A number of interactions between dipoles are possible because of the variety of possible orientations. Consider the dipole–dipole orientation of Fig. 3.6, in which the dipoles lie on a common axis with common direction. A net attraction is expected, because the negative tail of one dipole and the positive head of the second are the charges separated by the shortest distance. The distance between dipole centers is r, while each dipole has a charge separation of d. The two central charges are separated by a distance

$$r - (d/2) - (d/2) = r - d \qquad (3.55)$$

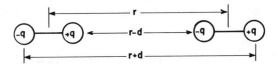

Fig. 3.6   Interactions between two dipoles aligned with their dipole axes colinear.

The two positive charges and the two negative charges are each separated by a distance r, while the two outermost charges have a separation

$$r + (d/2) + (d/2) = r + d \tag{3.56}$$

The total energy for all interacting charges is

$$U = [1/4\pi\varepsilon_0\varepsilon_r][-q^2/r - d) + (2q^2/r) + (-q^2/r + d)] \tag{3.57}$$
$$= -[q^2/4\pi\varepsilon_0\varepsilon_r r][1/(1 - d/r) - 2 + 1/(1 + d/r)] \tag{3.58}$$

Expanding each of the terms in (d/r) gives

$$U = -[q^2/4\pi\varepsilon_0\varepsilon_r r][1 + (d/r) + (d/r)^2 - 2 + 1 - (d/r) + (d/r)^2] \tag{3.59}$$
$$= -[q^2/4\pi\varepsilon_0\varepsilon_r r][2(d/r)^2] \tag{3.60}$$

The negative energy indicates a net attraction. The equation can be rearranged in the form

$$U = -[2(qd)(qd)/4\pi\varepsilon_0\varepsilon_r r^3] = -(2\mu_1\mu_2/4\pi\varepsilon_0\varepsilon_r r^3) \tag{3.61}$$

Just as a coulomb attraction was proportional to the charge on both ions, the dipole–dipole interaction is proportional to the product of the two dipole moments. However, because the effects of like-charge repulsion cancel some of the attractive energy, the energy is significant only at short range; this is reflected in the r^{-3} dependence of this equation.

Equation (3.61) gives the interaction energy for a single orientation for the two dipoles. Other orientations might also give a significant interaction. Consider the arrangement of Fig. 3.7. Both dipoles are now perpendicular to the length r, which connects their centers. One dipole is directed upward ($\theta_1 = 90°$), while the other is directed downward ($\theta_2 = 270°$). A net attraction is expected.

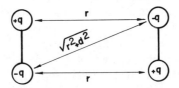

Fig. 3.7   Charge interactions for dipoles with parallel or antiparallel alignment.

The separation of charges with opposite sign is r. Charges of common sign are separated by a distance

$$[r^2 + d^2]^{1/2} \tag{3.62}$$

The total interaction energy is

$$U = [1/4\pi\varepsilon_o\varepsilon_r]\{(-2q^2/r) + [2q^2/(r^2 + d^2)^{1/2}]\} \tag{3.63}$$

An r is factored from each term, and the term

$$[1 + (d/r)^2]^{-1/2} \tag{3.64}$$

is expanded as

$$1 - \tfrac{1}{2}(d/r)^2 + \cdots \tag{3.65}$$

The expression becomes

$$U = -(q^2/4\pi\varepsilon_o\varepsilon_r r)[2 - 2 + (d/r)^2] \tag{3.66}$$
$$= -(q^2/4\pi\varepsilon_o\varepsilon_r r)(d/r)^2 \tag{3.67}$$
$$= -(\mu_1\mu_2/4\pi\varepsilon_o\varepsilon_r r^3) \tag{3.68}$$

The net interaction for this orientation is attractive but only one-half as large as that of the linear arrangement.

These specific results may be generalized to the general case where the dipoles each have some arbitrary orientation with respect to the interdipole axis. Dipole 1 has an angle θ_1 with respect to the interdipole axis, while dipole 2 has an angle θ_2. The expression involving these angles must produce a factor of 2 when the dipoles are colinear and a factor 1 when they are perpendicular to the interdipolar axis (that is, $\theta_1 = 90°$ and $\theta_2 = 270°$). For the factor of 2, a term of the form

$$2(\cos\theta_1)(\cos\theta_2) \tag{3.69}$$

is required. However, this term vanishes for $\theta = 90°$; the attraction between perpendicular dipoles suggests a second term of the form

$$-(\sin\theta_1)(\sin\theta_2) \tag{3.70}$$

where the minus sign produces a net attraction when the dipoles are anti-parallel.

When both dipoles lie in a common plane, the dipole–dipole interaction energy for arbitrary orientation will be

$$U = -\mu_1\mu_2(4\pi\varepsilon_o\varepsilon_r r^3)[2(\cos\theta_1)(\cos\theta_2) - (\sin\theta_1)(\sin\theta_2)] \tag{3.71}$$

The constant variables can again be collected and evaluated, and this energy can be expressed in kJ/mole as

$$U = -(6610d_1d_2/\varepsilon_r r^3)[2(\cos\theta_1)(\cos\theta_2) - (\sin\theta_1)(\sin\theta_2)] \tag{3.72}$$

where d_1, d_2, and r have units of angstroms.

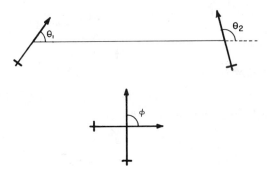

Fig. 3.8 (a) The orientation angles θ_1 and θ_2 for two interacting dipoles restricted to a plane. (b) The angle ϕ as a measure of relative rotation out of the plane.

The orientation factor for this expression has been simplified by restricting the two dipoles to a single plane. However, the dipoles may also rotate about the interdipolar axis; this addition angle of rotation ϕ must also enter the orientation factor. Let dipole 2 rotate out of the common plane by $\theta = 90°$. If $\theta_1 = \theta_2 = 90°$, this rotation still keeps both dipoles perpendicular to the interdipolar axis. The situation is clarified by looking along the interdipolar axis (Fig. 3.8). At $\phi = 90°$, the second dipole charges are equidistant from the dipole charges on dipole 1 and a net interaction energy of 0 is expected. If the rotation were continued so that $\phi_1 = 0°$ and $\phi_2 = 180°$, the interaction would move from a net repulsion to a net attraction, because the rotation places opposite charges adjacent. The ϕ-dependent function must then appear with the $(\sin \theta_1)(\sin \theta_2)$ term, and this is now

$$2(\cos \theta_1)(\cos \theta_2) - (\sin \theta_1)(\sin \theta_2)[\cos(\phi_1 - \phi_2)] \qquad (3.73)$$

where $\phi_1 - \phi_2 = \phi$ describes a net out-of-plane rotation.

Since the phospholipids contain dipoles, dipole–dipole interactions are possible when these molecules are part of a membrane. Because x-ray studies indicate a distance of 5 Å between chains, the distance between phospholipid molecules (and dipoles) is approximately 10 Å. If the charge separation in each molecule is 5 Å, then the maximal energy of interaction will be

$$U = -[2(1377)(5)(5)/10^3] = -6.89 \text{ kJ/mole} \qquad (3.74)$$

For this case, $\varepsilon_r = 1$, because no water molecules are expected between these closely spaced dipole molecules. This energy exceeds the thermal energy of 2.5 kJ/mole, so that such dipole–dipole interactions can stabilize the membrane at room temperature. Because the energy of rotation of a dipole in a large transmembrane electric field is approximately 2 kJ/mole, the dipoles are expected to maintain their stable configuration in the presence of such fields.

6. THE EFFECTS OF THERMAL AGITATION

The interaction between a charge and a dipole in solution was given by the equation

$$U = (q\mu \cos \theta)/4\pi\varepsilon_0\varepsilon_r r^2 \tag{3.75}$$

for a fixed value of θ. In solution, however, the charges will encounter the dipoles at all orientations and the interaction energy will be some averaged interaction energy over all these orientations. The averaging process for a dipole in solution was performed in Chapter 2; the average depended on the magnitude of the applied electric field and the temperature of the solution,

$$\langle\mu\rangle = \mu^2 E/3kT \tag{3.76}$$

For the case of a dipole–charge interaction, the field E that produces this average dipole will be the field generated by the charge,

$$E = (1/4\pi\varepsilon_0\varepsilon_r)(q/r^2) \tag{3.77}$$

Because the interaction energy of the average dipole in the field E is

$$U = -\bar{\mu}E = -[\mu^2/3kT]E^2 \tag{3.78}$$

the average interaction energy will be

$$U = -(1/4\pi\varepsilon_0\varepsilon_r)^2(\mu^2/3kT)(q/r^2)^2 \tag{3.79}$$

$$= (1/4\pi\varepsilon_0\varepsilon_r)^2(\mu^2 q^2/3kTr^4) \tag{3.80}$$

Since the thermal motions average the dipole, the resultant interaction energy is of very short range and depends on the inverse fourth power. This fourth-power dependence is generated by the E^2 dependence. One factor of E establishes an average dipole, while the second E consummates the interaction between the dipole and the charge.

Equation (3.80) can be reexpressed in kJ/mole,

$$U = -(3.66 \times 10^6)n^2\mu^2/r^4kT \tag{3.81}$$

where n is the number of charges on the ion, the distance r is in angstroms, and μ is in debyes.

The calculation of the average interaction between two dipoles will be more complicated than that of the charge–dipole interaction, because both dipoles can now assume various orientations. Despite this, the basic approach is similar. The average dipole for dipole 2 produced by the dipole field E_1 of dipole 1 is

$$\langle\mu_2\rangle = \mu_2^2 E_1/3kT \tag{3.82}$$

The electric field produced by dipole 1 has the form

$$E_1 = (A/4\pi\varepsilon_o\varepsilon_r)(\mu_1/r^3) \tag{3.83}$$

where the factor A contains the statistical information on the orientations of both dipoles. The average dipole–dipole interaction energy is

$$U = -\mu_2 E_1 \tag{3.84}$$

or

$$U = -(1/4\pi\varepsilon_o\varepsilon_r)^2 A^2 \mu_2^2 \mu_1^2 / 3kTr^6 \tag{3.85}$$

The inverse-sixth dependence arises from the squared electric field dependence.

7. INDUCED ELECTROSTATIC INTERACTIONS

A molecule may not have a permanent dipole moment. Although this might suggest that it cannot interact with an applied field, the molecule does consist of positive and negative components. An applied field might induce a charge separation in the molecule to produce a dipole moment that then interacts with the field. The situation is similar to that encountered when we considered thermal motions. The electric field from a charge produced an "averaged" dipole, which then interacted with the charge.

Obviously, the ability to form an induced dipole in a given molecule depends on the ability to separate opposite charges with the electric field. This distortability of the molecule is called its polarizability and is described with the parameter α. The average dipole moment per mole, P, produced when an electric field is applied to the molecule is then

$$P = \alpha E_{ext} \tag{3.86}$$

A similar expression is used to determine the average dipole induced in a given molecule,

$$\mu = \alpha' E_{ext} \tag{3.87}$$

where α' is a molecular, rather than a molar, quantity in this case.

The interaction energy between the induced dipole and the external electric field is

$$U = -\mu E \tag{3.88}$$

because the induced dipole is aligned with the field. The change in energy as the induced dipole changes is

$$dU = -E \, d\mu \tag{3.89}$$

or, using Eq. (3.87),

$$dU = -E(\alpha \, dE) \tag{3.90}$$

To determine the total energy for a given field, Eq. (3.90) must be integrated from zero field to the final field E,

$$U = -\tfrac{1}{2}\alpha E^2 \big|_0^E = -\tfrac{1}{2}\alpha E^2 \tag{3.91}$$

The E^2 dependence for this equation indicates that our equations will have inverse r dependences identical to those observed for the thermal agitation averages.

Consider a charge-induced dipole interaction. The electric field of the charge is

$$E = (1/4\pi\varepsilon_0 \varepsilon_r)(q/r^2) \tag{3.92}$$

and the interaction energy is

$$U = -\alpha E^2/2 = -(1/4\pi\varepsilon_0 \varepsilon_r)^2 \alpha[q^2/2r^4] \tag{3.93}$$

The equation can be simplified to

$$U = -690n^2\alpha/\varepsilon_r^2 r^4 \tag{3.94}$$

where n is the number of charges on the ion and α is the polarizability in units of 10^{-24} cm^3/molecule. If a water molecule has a polarizability of 1.5×10^{-24} cm^3/molecule, then a trivalent lanthanide ion at a distance $r = 6$ Å will produce a net interaction energy of

$$U = -(690)(9)(1.5)/6^4(80)^2 = 1.12 \text{ J/mol} \tag{3.95}$$

At this distance, the hydrated ion induces an interaction well below the thermal energy. If the ion loses its hydration sphere and approaches to a distance of 2 Å, the interaction energy increases to 582 kJ. Obviously, the dielectric constant of the water and the distance of separation play major roles in determining the magnitude of the interaction energy.

A dipole-induced dipole interaction utilizes the electric field produced by the permanent dipole,

$$E = (1/4\pi\varepsilon_0 \varepsilon_r)(\mu/r^3) \tag{3.96}$$

for $\cos(0) = 1$. The interaction energy is

$$U = -(1/4\pi\varepsilon_0 \varepsilon_r)^2 \alpha(\mu^2/2r^6) \tag{3.97}$$

To include the effects of various dipole orientations, the two dipoles must be averaged over all possible orientations so that, in general, this equation must also include an averaging factor A:

$$U = -(1/4\pi\varepsilon_o\varepsilon_r)^2 A\alpha(\mu^2/2r^6) \tag{3.98}$$

8. CAPACITANCE

In this chapter, we have used both the relative dielectric constant and the polarizability without regard to their experimental measurement. The key to such measurement is the molecule's ability to change capacitance when it is present as an insulating material. The capacitance measures the ability of a system to store charge. A typical parallel-plate capacitor is illustrated in Fig. 3.9. If a potential difference is applied between the plates, positive charge will flow to one plate and negative charge will flow to the other. The capacitance is a measure of the charge that can be stored per unit voltage,

$$C = q/V \tag{3.99}$$

with units of farads (C/V). For the parallel-plate capacitor, the capacitance is

$$C = \varepsilon_o A/d \tag{3.100}$$

when there is a vacuum between the plates.

For cylindrical cells, such as a nerve axon, the capacitance is related to the radii of the two plates (b outer, a inner). For a cylinder of length ℓ, the capacitance is

$$C = 2\pi\varepsilon_o\ell/[\ln(b/a)] \tag{3.101}$$

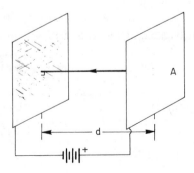

Fig. 3.9 Parallel-plate capacitor with plates of area A separated by a distance d.

In axonal systems, b is only slightly larger than a. Assume $b = a + d$ and expand the logarithm:

$$\ln(b/a) = \ln[(a + d)/a] = \ln[1 + (d/a)] = d/a \qquad (3.102)$$

The equation for capacitance is then

$$C = 2\pi\varepsilon_o \ell/(d/a) = 2\pi\varepsilon_o(\ell a)/d \qquad (3.103)$$

$$= [2\pi a\ell]\varepsilon_o/d = A\varepsilon_o/d \qquad (3.104)$$

The combination $2\pi a\ell$ is the surface area of the plates, so the equation reduces to that for a parallel-plate condenser.

If the vacuum between the plates is replaced by some insulating material, the capacitance of the parallel-plate combination will increase; the material enables the capacitor to hold more charge for the same voltage. It does so by balancing some of the electric field vector with a field created by the dipoles (or induced dipoles) in the material. This vector produced by the dipoles is the polarization vector. Note, however, that E and P have different units. Their units will balance if E is multiplied by the permittivity, ε_o.

$$P \text{ (dipole moment/volume)} = P \text{ (C m/m}^3) = P \text{ (C/m}^2) \qquad (3.105)$$

$$\varepsilon_o(C^2/N\ m^2)E(N/C) = \varepsilon_o E(C/m^2) \qquad (3.106)$$

The total charge in a unit area will now be the sum of the charge associated with the field in vacuum plus the charge created by the dipoles,

$$\varepsilon_o E + P \qquad (3.107)$$

This total charge is used to define a new parameter, the electric displacement D:

$$D = \varepsilon_o E + P \qquad (3.108)$$

which is also defined as

$$D = \varepsilon_r \varepsilon_o E \qquad (3.109)$$

Thus, the relative dielectric constant ε_r provides a measure of the additional charge separation generation in the medium as shown by rearranging Eqs. (3.108) and (3.109).

$$P = D - \varepsilon_o E = (\varepsilon_r - 1)\varepsilon_o E \qquad (3.110)$$

The relative dielectric constant can be determined by measuring capacitance both with and without the dielectric medium. This factor, in turn, gives a measure of the polarization, that is, dipole moment per unit volume. From this information, detailed information on dipole moments and induced dipole moments is determined.

9. IMAGE CHARGES

Charge distributions such as dipoles produce complicated expressions for the potential, but these potentials can be calculated by summing the potentials produced by the individual charges. However, consider the configuration in Fig. 3.10, in which a charge is placed near a large, flat, conducting plate. Since we seek relative potential differences, the potential of the plate is selected as zero, that is, ground.

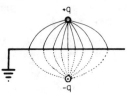

Fig. 3.10 Image charge used to determine the electric lines of force between a charge and a grounded plate.

This problem can be solved by finding a different arrangement of charge that produces the same potential. The potential must disappear at the plate. To cancel the potential generated by the positive charge at the plate, a second negative charge is placed an equal distance behind the plate—that is, a mirror image. The combination of the charge and its image charge is now identical to the problem of the single charge and plate. The two-charge problem is essentially identical to the dipole problem considered earlier, and those solutions can be used here. However, for the plate-charge problem, the electric field is found only to the right of the plate. Instead of an image charge, a charge of $-q$ will be drawn onto the plate. The electric field lines of force are shown in Fig. 3.10.

The image charge technique can also be used for two grounded plates that meet at right angles. A charge is placed between these plates, as shown in Fig. 3.11. Each plate is then replaced by an image charge of $-q$. However,

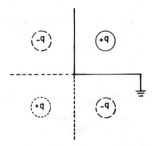

Fig. 3.11 Image charge required to describe electric lines of force between a charge and perpendicular grounded plates.

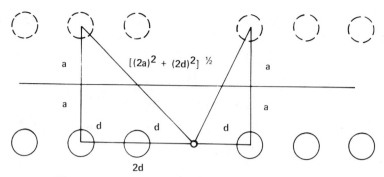

$$[(2a)^2 + (2d)^2]^{1/2}$$

Fig. 3.12 Image charges for one charge in a line of charges.

there would then be two image charges for the single positive charge. A third, positive, image charge is added diagonally opposite the original positive charge as shown. This set of four charges will produce fields identical to that of the two plates and single charge. Once again, the electric field will only be found in the original quadrant for the plate-charge system.

Both these image charge systems are simpler than those that would be found on a membrane system. In this case the membrane might be confronted with a large number of distinct charges, and it becomes necessary to develop a set of image charges consistent with this array. Consider the simple "one-dimensional" membrane in Fig. 3.12. Charges are arranged in a line parallel to the surface, with a separation d, and the line is a distance a from the surface. Consider a single charge in the chain; it will experience a repulsion from the adjacent charges at distance d. However, the charges at $2d$, $3d$, and so on, will also produce a net repulsion, although their contribution will become progressively smaller with increasing distance.

In order to establish ground potential for the surface, a set of image charges a distance a behind the surface is required. These charges will have opposite sign, so there is a net attraction to the charge of interest. The image charges will lie at a distance

$$[(nd)^2 + (2a)^2]^{1/2} \tag{3.111}$$

from the charge of interest. Note that we do not include an image charge for the ion under examination. For this calculation, that ion is not present; the potential at the reference ion position is calculated from the line of charge and the surface.

Let

$$B = 1/4\pi\varepsilon_0\varepsilon_r \tag{3.112}$$

The effects of different ε_r on either side of the membrane will be considered below. The charges in the line now sum to give a net repulsive potential of

$$V_{rep} = B(2q/d) \sum (1/n) \tag{3.113}$$

because nd is the distance to the nth charge. The factor of 2 indicates that there are charges on either side.

The image charges will exert a net attraction,

$$V_{att} = \sum - B(2q)/[(2a)^2 + (nd)^2]^{1/2} \tag{3.114}$$

$$= \sum - B(2q/nd)/[1 + (2a/nd)^2]^{1/2} \tag{3.115}$$

The denominator is expanded as

$$[1 + (2a/nd)^2]^{1/2} = 1 - \tfrac{1}{2}(2a/nd)^2$$

and

$$V_{att} = \sum - B(2q/nd)[1 - \tfrac{1}{2}(2a/nd)^2] \tag{3.116}$$

Combining both potentials and summing over all values of n gives

$$V_{tot} = B[2q/d] \sum [(1/n) - (1/n) + (2a^2/d^2n^3)] \tag{3.117}$$

$$= \sum B(4qa^2/d^3n^3) = (4.8qa^2/d^3)B \tag{3.118}$$

The net repulsion energy experienced by the reference charge, $+q$, is

$$U = 4.8Bq^2a^2/d^3 \tag{3.119}$$

The extension of two-dimensional arrays proceeds in a similar manner. The adjacent charges and incorporated into rings around the reference ion, and the ring potential and image ring potential are summed.

What effect will media of different dielectric constant on each side of the plate have on image charge calculations? Consider the case of a single charge in medium 1 (ε_{r1}) facing a plate with a second medium of different dielectric constant (ε_{r2}) behind it. Because of these different dielectrics, the charges q_1 and q_2 used to set up the image problem must be modified. The charge in Region 2 is

$$q_2 = q[(\varepsilon_{r1} - \varepsilon_{r2})/(\varepsilon_{r1} + \varepsilon_{r2})] \tag{3.120}$$

while the charge for Region 1 is

$$q_1 = q[2\varepsilon_{r2}/(\varepsilon_{r1} + \varepsilon_{r2})] \tag{3.121}$$

PROBLEMS

1. Show that the units of the permittivity ε_0 are farads per meter.

2. The magnitude of an electric field will be proportional to the number of lines passing through a surface. Sketch the fields for ions with $z = 1$, 2, and 3 within a spherical surface. Sketch the lines of equal potential for these ions.

3. For smaller values of r, the linearization approximation (Taylor expansion) breaks down. Compare the dipole potential calculated using this expansion with the exact result for the case where $d = r = 5$ Å and $\theta = 0°$.

4. A simple quadrapole consists of a central divalent charge flanked with two monovalent charges separated from it by a distance d. If the three charges form a linear array, determine the region(s) of maximum and minimum potential.

5. Develop a potential for the distance r and angle $\theta = 0°$ for the quadrapole in Problem 4; $\theta = 0°$ lies on the linear axis. Calculate the potential for $\theta = 90°$.

6. A dipole is aligned in the x direction. The field in the x direction is not constant but increases with increasing x, that is, $dE/dx = $ constant. The negative charge will then experience a different force than the positive charge, and the dipole will translate in the field. Determine this force for a dipole of length d.

7. Develop the general expression for the potential of a dipole for arbitrary r and θ using the law of cosines.

8. A dipole can only have two configurations ($\theta = 0°$ or $180°$) with respect to a field E. Determine the temperature at which 75% of the dipoles will be aligned with the field as a function of E.

Chapter 4

Solution and Surface Thermodynamics

1. SOLUTION CHEMICAL POTENTIALS

Expressions for work, energy, and other thermodynamic parameters can be derived for gaseous systems by using gaseous equations of state such as the ideal gas law. For example, the Maxwell relation

$$(\partial S/\partial V)_T = (\partial P/\partial T)_V \tag{4.1}$$

permits a determination of the entropy change once the variation of P with T is determined from the equation of state. Calculations of this type for solutions are more difficult, because there is no convenient equation of state. However, two solutions with a solute at different concentrations would be expected to have different free energies, because the solute will diffuse toward the more dilute solution. The driving free energy must now be expressed in some quantitative manner.

Free energies for solutions can be developed via an indirect approach. At equilibrium, the solution must have the same chemical potential as the vapor above the solution. The free energy of the vapor is determined using the relationships developed for gas-phase systems. Because this vapor and the solution must have the same chemical potential at equilibrium, the solution chemical potential is also known, that is,

$$\mu_{solution} = \mu_{vapor} \tag{4.2}$$

Any change in the chemical potential of the vapor must produce an equal change in the solution chemical potential to maintain equilibrium:

$$d\mu_{solution} = d\mu_{vapor} \tag{4.3}$$

If the vapor is described by the ideal gas equation of state, the chemical potential at some constant temperature T is

$$d\mu = \overline{V}\, dP = RT\, d(\ln P) \tag{4.4}$$

where $\overline{V}$ is the partial molar volume of the vapor, equal to RT/P for an ideal gas. The chemical potential can be expressed in joules by using $R = 8.31$ J/K mol.

Although many common solvents such as water have vapors that may deviate from the ideal gas law, the ideal gas law provides a reasonable starting point for a definition of the chemical potential of a vapor. More complicated equations of state would alter the functional form of $\overline{V}$, but the resultant development of the equations would parallel those shown here. For such nonideal vapors, Eq. (4.4) is maintained but the pressure is modified to reflect the nonideality of the vapor. This modified pressure is called the fugacity, f, and the equation has the form

$$d\mu = RT\, d(\ln f) \tag{4.5}$$

To determine the energy required to change the vapor pressure from some initial pressure P_i to some final pressure, P_f, Eq. (4.4) is integrated through a series of pressures bracketed by these pressures, so that

$$\int_{\mu_i}^{\mu_f} d\mu = RT \int_{P_i}^{P_f} dP/P \tag{4.6}$$

$$\Delta\mu = \mu_f - \mu_i = RT \ln(P_f/P_i) \tag{4.7}$$

The equation becomes more convenient if a specific pressure is selected as a reference pressure so that all chemical potential differences are determined relative to this pressure. For example, the reference pressure could be chosen as the vapor pressure of pure water, p°, at the temperature of interest, and the equation could be rewritten as

$$\mu_f = \mu^\circ + RT \ln(P_f/p^\circ) \tag{4.8}$$

Equation (4.8) becomes extremely useful when we realize that nonvolatile solutes added to the solvent will lower the vapor pressure of that solvent. Thus, there is a connection between the concentration of solute and the free energy observed for the vapor. A number of linear relationships of this sort exist. One of the more common is Raoult's law, in which the mole fraction of a volatile species determines the vapor pressure of that species,

$$P_j = X_j p_j^\circ \tag{4.9}$$

Raoult's law defines an ideal solution. Unlike the ideal gas model, in which there are no interactions, interactions are now permitted, but these interactions must be the same for all species—that is, solvent–solvent interaction

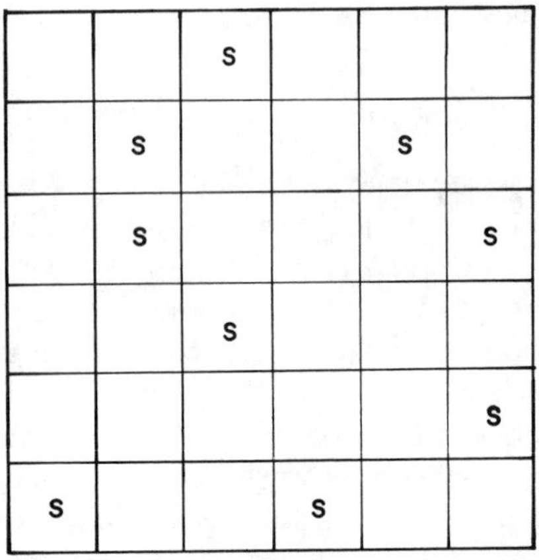

Fig. 4.1 A simple cell model of the surface of a solution.

energies are identical to solvent–solute interaction energies in the ideal solution. The molecules in solution will always experience the same local environment for any proportion of solvent and solute.

A rationale for Raoult's law can be developed by considering only the distribution of molecules on the surface. In Fig. 4.1, the surface is approximated as an array of 36 squares. If the solvent mole fraction is 0.75, 27 of these squares will contain solvent and the remaining 9 will contain solute. In the pure solvent, molecules could escape from all 36 squares; in the solution, the squares available for escape drop to 27. The ratio of the pressures will be proportional to the squares available, that is,

$$27/36 = 0.75 = P_j/p_j^o \tag{4.10}$$

Using Raoult's law, we can now calculate the chemical potential change of the vapor when solute is added to the pure solvent,

$$\mu_f = \mu^o + RT \ln(P_j/p_j^o) = \mu^o + RT \ln(X_j p_j^o/p_j^o)$$

$$\mu_f = \mu^o + RT \ln X_j \tag{4.11}$$

Because the vapor and solution phases are in equilibrium, Eq. (4.11) is also the chemical potential for the solution,

$$\mu_{\text{solution}} = \mu_{\text{vapor}} = \mu^o + RT \ln X_j \tag{4.12}$$

The reference state μ^o is for pure solvent. However, other reference states—such as 1 atm—could also be selected. The energy necessary to change the pressure from 1 atm to the vapor pressure,

$$RT \ln(p^o/1) \tag{4.13}$$

would then be added to Eq. (4.12).

A variety of reference states could be chosen. However, in most applications, we will be considering the effects of changes in chemical potential (or concentration); a common reference state will exist and will cancel when such differential changes are considered. This is shown clearly when we take the differential of both sides of Eq. (4.12):

$$d\mu_{solution} = d\mu_{vapor} = RT \, d(\ln X) \tag{4.14}$$

When this differential is integrated between two limits in the examples that follow, the reference chemical potential will not enter the calculations, because chemical potential differences will be calculated.

Raoult's law produces an expression for chemical potential that depends on the natural logarithm of the solvent. However, it is often convenient to use the amount of solute as the variable of interest. In dilute binary solutions, the logarithm of the mole fraction of solvent can be converted to the mole fraction of solute by expanding the logarithm in a Taylor series:

$$\ln(X_{solvent}) = \ln[1 - X_{solute}] = -X_{solute} \tag{4.15}$$

For example, $\ln(0.9) = -0.105$, whereas the linear approximation gives -0.1. The elimination of the logarithm makes this a convenient approximation.

When a gas is dissolved in water, the gas pressure and the concentration of gas in solution are also linearly related,

$$p = kc \tag{4.16}$$

where c is the molar concentration of gas in solution. By selecting a 1 M concentration of the gas as a standard state, the free energy for the gas in solution is

$$\mu_f = \mu^o + RT \, \ln[(kc)/(k \cdot 1 \text{ M})] = \mu^o + RT \, \ln(c) \tag{4.17}$$

Equation (4.17) suggests that the logarithm of any concentration variable will provide a measure of the chemical potential.

Although Eq. (4.17) was derived for a gas in solution, we extend its use to include any species that is added to solvent, even if this species does not have a vapor pressure. Equation (4.17) can be used with ions in solution, for example. The differential form for use in equilibrium studies is

$$d\mu = RT \, d(\ln c) \tag{4.18}$$

This differential form will become particularly important when we consider multiphase (membrane) systems. If the free energy or chemical potential in one phase changes by a differential amount, the free energy or chemical potential in the second phase must change by an identical amount to maintain equilibrium between the two phases.

Although Eq. (4.18) was developed for the solute in a binary system, the relationship will hold for each species that has an independent existence in solution. When sodium chloride is dissolved in water, chemical potential relations can be written for both the Cl^- and Na^+. The condition of electroneutrality will provide an additional relationship between the two independent expressions.

The independence of the two ionic chemical potentials is based on the assumption that all the ions are independent in solution. Although this is a reasonable assumption for very low concentrations, the probability of interaction increases as the concentrations of the ions in solution increase. Under such circumstances, Eq. (4.18) constitutes a first approximation to the actual chemical potential. In order to preserve the form of Eq. (4.18), the concentration c is replaced by the activity a when interionic interactions become significant, that is,

$$d\mu = RT \, d(\ln a) \qquad (4.19)$$

The activity a is related to the actual ion concentration via the linear relationship

$$a = \gamma c \qquad (4.20)$$

where the proportionality constant γ is called the activity coefficient.

In this chapter, concentration will be used in derivations; the derivations for activity are identical. For membrane systems, the activity coefficients will enter as a ratio. Therefore, some effects of this interactive behavior will cancel and reduce the difference between the ideal and nonideal equations for dilute solutions.

2. THE OSMOTIC PRESSURE

To illustrate the use of the chemical potential, consider the case where water, rather than solute, equilibrates between the two sides of the membrane. To establish equilibrium, water must be free to move through the membrane. Initially, assume that no solute molecules or ions are membrane permeable. The problem can then be restricted to a comparison of the chemical potentials of water.

One major advantage of thermodynamics is the fact that the molecular details of the membrane are not considered. Thermodynamics compares only the chemical potentials of the two solutions that the membrane separates. Anything that permitted a flow of water could be inserted between the two phases and the chemical potentials would still balance to define equilibrium. For example, the "membrane" might be 2 m thick, and water would obviously require a considerable time to flow between the two phases. However, thermodynamics is not concerned with time; it simply states that, *at equilibrium*, the chemical potentials of both phases will be equal.

Because details of the membrane structure do not have to be included, the problem reduces to comparing changes in the chemical potentials of both solution phases,

$$d\mu_1 = d\mu_2 \tag{4.21}$$

The chemical potentials will include a summation of the different thermodynamic variables that can contribute to a flow of water through the membrane. The concentration is the obvious first choice. Water will flow from a dilute salt solution to a more concentrated one; the dilute solution then becomes more concentrated, while the concentrated solution is diluted. However, the solute concentrations are not used in this case, because solute cannot flow through the membrane. The chemical potential of water must be considered, and it is based on the mole fraction of water in the phase [Eq. (4.14)],

$$d\mu_1 = RT\, d(\ln X_1) \tag{4.22}$$

or, for phase 2,

$$d\mu_2 = RT\, d(\ln X_2) \tag{4.23}$$

If the mole fractions of water on both sides were the same, there would be no preferred direction of flow and the system would be at equilibrium. If the two mole fractions are different, it becomes quite difficult to balance the system. Consider the simple case where $X_1 = 1$, for pure water, and $X_2 \neq 1$, such that solute is present on Side 2. The common standard chemical potential would cancel from both sides, and the chemical potential on Side 1 would be zero because $\ln 1 = 0$. Because the mole fraction on Side 2 is fractional, its chemical potential will be negative, and the water flow must commence from Side 1 with its higher chemical potential to Side 2. The flow would continue until both sides attained the same concentration. Of course, this is impossible; the solution on Side 2 might become extremely dilute after sufficient water had flowed, but it would never be pure water. In order to establish a viable chemical equilibrium, a second chemical potential term is required to counterbalance the flow.

For this system, the weight of the water that flows can be used to counterbalance the drive toward equal concentrations on each side. As water flows from Side 1 to Side 2, the level of the solution will begin to rise. The resultant pressure head will force water to flow back toward Side 1. Eventually, the

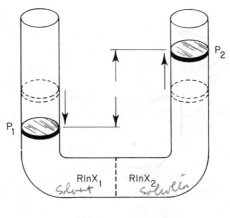

Fig. 4.2 An osmotic pressure gradient produced through a semipermeable membrane when $X_2 > X_1$.

system will reach an equilibrium in which the driving force of the concentration gradient is just balanced by the driving force of the pressure head. The situation is illustrated in Fig. 2.1. The equilibrium pressure difference between the two phases is the osmotic pressure, Π.

To introduce the chemical potential contribution of the pressure difference, consider the original expression for the free energy,

$$dG = -S\,dT + V\,dP \qquad (4.24)$$

where dG is the free energy change for the solution. The chemical potential is

$$\mu_{H_2O} = (\partial G/\partial n_{H_2O})_{n_{solute}} \qquad (4.25)$$

For these more dilute solutions, the chemical potential can be approximated to be the free energy divided by the moles of *solution*, n_t,

$$d\mu = dG/n_t = -(S/n_t)\,dT + (V/n_t)\,dP \qquad (4.26)$$

For constant temperature,

$$d\mu = \overline{V}\,dP \qquad (4.27)$$

For dilute solutions, $\overline{V}$ can be approximated as the volume of solution per mole of solution, that is, addition of small amounts of solute has a minimal effect on the volume of the solution.

If solutions are present on both sides of the membrane, both must be capable of developing a pressure gradient, so Eq. (4.27) must be added to the chemical potential expression. The expression is now

$$RT\,d(\ln X_1) + \overline{V}\,dP_1 = RT\,d(\ln X_2) + \overline{V}_2\,dP_2 \qquad (4.28)$$

For this equation, dilute solutions are preferred because differences in the partial molar volumes are small and they can be equated to a common partial molar volume, $\overline{V}$.

In order to develop an expression for the pressure difference in terms of the mole fractions, X_i, Eq. (4.28) must be integrated through a series of equilibrium configurations until it reaches the final equilibrium configuration of interest. If both concentrations are the same, then no pressure gradient is required. An ideal choice of initial state is then

$$X = X_1 = X_2 \qquad P = P_1 = P_2 \tag{4.29}$$

The system is then integrated from X and P to the final mole fractions (X_1 and X_2) and final pressures (P_1 and P_2),

$$\int_X^{X_1} RT\, d(\ln X_1) + \int_P^{P_1} \overline{V}\, dP_1 = \int_X^{X_2} RT\, d(\ln X_2) + \int_P^{P_2} \overline{V}\, dP_2 \tag{4.30}$$

and the final result is

$$RT[\ln(X_1/X)] + \overline{V}(P_1 - P) = RT[\ln(X_2/X)] + \overline{V}(P_2 - P) \tag{4.31}$$

Cancelling terms common to both sides yields

$$RT(\ln X_1) + \overline{V}P_1 = RT(\ln X_2) + \overline{V}P_2 \tag{4.32}$$

or

$$\overline{V}\Pi = \overline{V}(P_2 - P_1) = RT(\ln X_1 - \ln X_2) \tag{4.33}$$

where Π is the osmotic pressure difference.

Although Eq. (4.33) is perfectly adequate, it can be recast more simply by expanding the logarithms in their Taylor expansions, because this is justified for dilute solutions. Therefore,

$$\ln X_{H_2O} = \ln(1 - X_{solute}) = -X_{solute} \tag{4.34}$$

Inserting this modification into Eq. (4.33) gives

$$\overline{V}\Pi = RT(X_{s2} - X_{s1}) \tag{4.35}$$

The osmotic pressure is directly proportional to the difference in solute concentrations in the two solutions. The equation can be simplified still further by noting that

$$X_{s2} = n_{2s}/n_t \qquad X_{s1} = n_{1s}/n_{1t} \tag{4.36}$$

Therefore,

$$\Pi = RT[(n_{2s}/n_t \overline{V}) - (n_{1s}/n_t \overline{V})] \tag{4.37}$$

Because $n_t \bar{V}$ is approximately the volume V of the dilute solution and n_{2s}/V is the molar concentration, c_2, the expression becomes

$$\Pi = RT(c_2 - c_1) = RT\, \Delta c \qquad (4.38)$$

Osmotic pressure studies provided one of the earliest indications of salt dissociation in solutions. The concentrations in Eq. (4.38) include everything except water. Thus, if a 1 M solution of NaCl is used, there actually is 2 M solute contributing to the decrease in the water chemical potential. If a salt breaks into i individual ions in solution, then the equation must be rewritten as

$$\Pi = iRT(c_2 - c_1) \qquad (4.39)$$

where the c_i terms are the molar concentrations of undissociated salt.

The equation may also be used for partial ionic dissociation. For example, if a fraction x of the molecules of formula $A_n B_m$ dissociate, the initial concentration of undissociated salt is reduced to

$$c(1 - x) \qquad (4.40)$$

while

$$(m + n)cx \qquad (4.41)$$

free ions are formed. The total concentration of all solute particles in solution is

$$c(1 - x) + (m + n)cx = c[1 + (m + n - 1)x] \qquad (4.42)$$

Although we have derived an expression for the osmotic pressure, we have given no indication of the magnitudes of the parameters involved. Consider nondissociating solutes of concentrations 1 M and 0.5 M, respectively. The osmotic pressure is

$$\Pi = (0.082 \text{ liter atm/K} \cdot \text{mol})(298 \text{ K})(1\ M - 0.5\ M) = 12.2 \text{ atm} \qquad (4.43)$$

The pressure difference required to establish equilibrium is huge. Each atmosphere will support 33 ft of water, so the water column necessary to counterbalance the concentration driving potential will be 403 ft high. Most biological systems could not provide sufficient water to maintain such a pressure gradient.

In order to circumvent this problem, osmotic pressure is usually measured by applying an external pressure just sufficient to prevent a water flow. The resultant equilibrium is the same, but a flow of water through the membrane is not required.

3. THE DONNAN EQUILIBRIUM

The osmotic pressure of the previous section was developed on the assumption that only solvent could communicate between the two phases via the membrane. Additional thermodynamic effects are possible for a membrane is which only the solute molecules or ions are free to pass through the membrane. For nonionizable molecules, the chemical potential will be

$$d\mu = RT \, d(\ln c) \qquad (4.44)$$

where c is again the molar concentration of the solute. Except for the change in concentration units, this form parallels that for the solvent. For ionic systems, however, the problem becomes more interesting. Consider a simple 1:1 electrolyte, such as KCl, that dissociates completely in solution. For a concentration c_1, there will be equal concentrations of K^+ and Cl^- ions in solution. The individual chemical potentials are

$$d\mu_1(K^+) = RT \, d[\ln c_1(K^+)] \qquad (4.45)$$

and

$$d\mu_1(Cl^-) = RT \, d[\ln c_1(Cl^-)] \qquad (4.46)$$

If the membrane is permeable to both cations and anions, then any cations that pass through the membrane should be accompanied by an anion to preserve electroneutrality. The chemical potentials cannot be considered individually under such circumstances. For this reason, they are added to produce a net chemical potential. The chemical potential for Side 1 is then

$$d\mu_1 = RT \, d[\ln c_1(K)] + RT \, d[\ln c_1(Cl)]$$
$$= RT \, d\{\ln[c_1(K)c_1(Cl)]\} \qquad (4.47)$$

The chemical potential for Side 2 will be

$$d\mu_2 = RT \, d\{\ln[c_2(K)c_2(Cl)]\} \qquad (4.48)$$

At equilibrium, these chemical potential changes must be equal. Integrating the two equations starting from the same initial concentrations produces

$$RT\{\ln[c_1(K)c_1(Cl)]\} = RT\{\ln[c_2(K)c_2(Cl)]\} \qquad (4.49)$$

The RT and the logarithm can be eliminated from each side to give

$$c_1(K)c_1(Cl) = c_2(K)c_2(Cl) \qquad (4.50)$$

The product of the concentrations is central to the chemical potential balance problem for ions. Each product has the form of an equilibrium constant, that is, solubility product. In this case, however, the equilibrium constants for each phase are being balanced against each other.

At first glance, Eq. (4.50) appears to contain very little information. Because the cation and anion concentrations on each side should be equal, there is little flexibility in the equation: thus $c_1 = c_1(K) = c_1(Cl)$, and

$$c_1^2 = c_2^2 \quad \text{or} \quad c_1 = c_2 \tag{4.51}$$

The equation simply states that the concentrations must be equal at equilibrium.

Equation (4.50) can be used to calculate the approach to equilibrium from a nonequilibrium situation if the volumes of both baths are the same. For example, if the concentrations of KCl on Sides 1 and 2 are 1 M and 2 M respectively, an amount x will be lost from Side 2 and gained by Side 1, that is,

$$(1 + x)^2 = (2 - x)^2$$
$$x = 0.5 \tag{4.52}$$

The concentrations on both sides are 1.5 M.

Now consider the alternative case in which the concentrations remain the same (1 M and 2 M on Sides 1 and 2), but the volume for Side 1 is 1 liter while the volume for Side 2 is 0.5 liter. There are equal numbers of ions on each side of the membrane in this case, even though the concentrations are different. Because the concentrations are different, there will be a flow of ions toward Side 1. The problem appears identical to the previous problem, because the concentrations are identical. The equation would predict concentrations of 1.5 M on each side. A moment's reflection shows that there are not enough ions to establish this concentration on each side. The total number of ions on both sides is

$$(1 \ M)(1 \text{ liter}) + (2 \ M)(0.5 \text{ liter}) = 2 \text{ mol} \tag{4.53}$$

while the total volume is 1.5 liters. Thus, the concentration expected when both solutions become completely homogeneous is

$$2 \text{ mol}/1.5 \text{ liters} = 1.33 \ M \tag{4.54}$$

The system "runs out" of ions before it can reach the equilibrium required by Eq. (4.50). The concentrations on both sides will be equal, but smaller than predicted. For this reason, Eq. (4.50) must be used with care.

The calculations implicit in Eq. (4.50) may appear rather trivial at this point, because they appear to be nothing more than a formalized way to determine that solute concentrations on either side of a membrane permeable to them must be equal. However, Eq. (4.50) is more general than this. The equation requires no information about the source of the ions on each side. If we could artificially increase the concentration of K^+ on Side 1 without increasing the Cl^- concentration, then both K^+ and Cl^- concentrations on this side would have to decrease to maintain Eq. (4.50) and the condition of electroneutrality. The problem then parallels common ion effects in equilibrium studies.

Although it may appear difficult to create an excess of K^+ ions relative to Cl^- ion, it is really quite simple. Extra K^+ is added as the salt of some large, membrane-impermeable anion, such as a protein molecule. Since this A^- anion cannot cross the membrane, it cannot be included in the chemical potential calculations and acts merely as a spectator ion. The K^+ ion added in this manner contributes to the total K^+ concentration that must be used in the equilibrium calculations. Because there is now excess K^+ on one side of the membrane, it will flow to the second side while carrying along Cl^- anions to maintain electroneutrality. Thus, the net concentration of KCl on the second side (Side 2) will increase relative to the system when no KA is present.

To illustrate this phenomenon, consider the simple case in which the system is at equilibrium with $1\ M$ concentrations of KCl on each side. Now $1\ M$ KA is added to Side 1. Because of this perturbation, a concentration x of KCl will flow to Side 2. We can tabulate the situation as

	Before equilibrium	At equilibrium
Side 1	$2\ M\ K^+$	$(2\ M - x)K^+$
	$1\ M\ Cl^-$	$(1\ M - x)Cl^-$
	$1\ M\ A^-$	$1\ M\ A^-$
Side 2	$1\ M\ K^+$	$(1\ M + x)K^+$
	$1\ M\ Cl^-$	$(1\ M + x)Cl^-$

The equilibrium values are substituted into Eq. (4.50) to give

$$(2 - x)(1 - x) = (1 + x)(1 + x)$$
$$2 - 3x + x^2 = 1 + 2x + x^2$$
$$5x = 1$$
$$x = 0.2 \qquad (4.55)$$

so that the final concentrations are

Side 1: $1.8\ M\ K^+, 0.8\ M\ Cl^-, 1\ M\ A^-$

Side 2: $1.2\ M\ K^+, 1.2\ M\ Cl^-$ (4.56)

Both sides are still electroneutral. The products of cation and anion concentrations are identical at equilibrium,

$$(1.8)(0.8) = 1.44 = (1.2)(1.2) \qquad (4.57)$$

4. DONNAN EQUILIBRIA FOR POLYVALENT IONS

When the two bathing solutions in a membrane system contain polyvalent ions, the basic equilibrium relation must be modified accordingly, because the concentrations of anions and cations may be different. The basic derivation parallels that for the $1:1$ electrolytes. Consider the $2:1$ electrolyte, $CaCl_2$. If this salt dissociates completely, then there will be twice as many Cl^- ions in solution. Since the free energy is based on 1 mole of ions, the chemical potential contribution for the Cl^- must be doubled. The equation for the salt on Side i is now

$$d\mu_i = RT \, d \ln[c_i(Ca^{2+})] + 2RT \, d \ln[c_i(Cl^-)] \tag{4.58}$$

Combining the two terms gives

$$d\mu_i = RT \, d \ln[c_i(Ca^{2+})c_i(Cl^-)^2] \tag{4.59}$$

The chemical potentials for each bathing solution must be equal at equilibrium,

$$RT \ln[c_1(Ca^{2+})c_1(Cl^-)^2] = RT \ln[c_2(Ca^{2+})c_2(Cl^-)^2] \tag{4.60}$$

Eliminating RT and the logarithm yields

$$c_1(Ca^{2+})c_1(Cl^-)^2 = c_2(Ca^{2+})c_2(Cl^-)^2 \tag{4.61}$$

To illustrate the use of this equation, consider a system in which the concentrations of Ca^{2+} on each side of the membrane are initially 1 mM each. For electroneutrality, the concentrations of Cl^- on each side must be 2 mM. Now 1 mM CaA, with A^{2-} an impermeable anion, is added to Side 2. This salt will induce the loss of x mM Ca^{2+} and $2x$ mM Cl^- from Side 2. The changes can be tabulated as

	Before equilibrium	After equilibrium
Side 1	1 mM Ca^{2+}	$(1 + x)$ mM Ca^{2+}
	2 mM Cl^-	$(2 + 2x)$ mM Cl^-
	0 mM A^{2-}	0 mM A^{2-}
Side 2	2 mM Ca^{2+}	$(2 - x)$/mM Ca^{2+}
	2 mM Cl^-	$(2 - 2x)$/mM Cl^{-1}
	1 mM A^{2-}	1 mM A^- (4.62)

The Donnan equilibrium is then

$$(1 + x)(2 + 2x)^2 = (2 - x)(2 - 2x)^2$$
$$(1 + x)^3 = (2 - x)(1 - x)^2$$
$$1 + 3x + 3x^2 + x^3 = 2 - 5x + 4x^2 - x^3$$
$$2x^3 - x^2 + 8x - 1 = 0 \tag{4.63}$$

Because the cubic terms have opposite signs, there are no convenient cancellations to simplify this equation. A programmable calculator could be used to determine the roots of this equation. Alternatively, the simple approximations that are used for equilibria can be used to determine the value of x. Use the linear portion of the equation to determine x to a first approximation, that is,

$$8x = 1 \qquad x = 0.125 \qquad (4.64)$$

Because the negative squared term will tend to reduce this value, round 0.125 to 0.13 and use this as a first approximation to complement 0.125. The two results give

$$(0.13) \qquad 0.0044 - 0.0169 + 1.04 - 1 = 0.027$$

$$(0.125) \qquad 0.0039 - 0.0156 + 1 - 1 = -0.012 \qquad (4.65)$$

Continuing in this manner, the value of x is found to be about 0.127. The equilibrium concentrations are then

Side 1:   1.127 mM Ca^{2+} 2.254 mM Cl^-

Side 2:   1.873 mM Ca^{2+} 1.746 mM Cl^- 1 mM A^{2-} (4.66)

Both sides are electroneutral. The equilibrium products are

$$(1.127)(2.254)^2 = 5.726 \qquad \text{and} \qquad (1.873)(1.746)^2 = 5.709 \qquad (4.67)$$

The small difference in the two products is due to roundoff error in x.

5. THE ELECTROCHEMICAL POTENTIAL

To develop the Donnan equilibria, we chose a membrane permeable to both anions and cations. If an excess of one ion was created on one side of the membrane, it would induce a flow of an equal number of anions to preserve electroneutrality on both sides. Considerable energy would be required to separate the positive and negative charges, so both must flow.

If the membrane were permeable only to cations, the situation would be different. The chemical potential would involve only the cation concentration. This potential energy might then be able to produce a charge separation and its concomitant electrical potential. Once again, the system will reach a balance point. In this case, the energy of the concentration gradient will be opposed by the electrical potential, which seeks to recombine the separated charges.

In order to study this problem, a chemical potential contribution from the electrical potential is required. The electrical potential ψ has the units of joules per coulomb. The electrical potential is then the work that could be done when 1 C of charge is moved across the potential difference. For the movement of dq of charge, the energy change is

$$W = \psi \, dq \tag{4.68}$$

Because the system will do work when the potential is positive, this term is included in the differential expression of the first law as a negative,

$$dU = T \, dS - P \, dV - \psi \, dq \tag{4.69}$$

The first two terms on the right-hand side can now be Legendre-transformed to produce a Gibbs free energy,

$$dG = -S \, dT + V \, dP - \psi \, dq \tag{4.70}$$

A major reason for the choice of free energy rather than internal energy lies in the fact that the intensive variables T and P are the variables in differential form. Such variables are easy to control because they are not dependent on the quantity of material. The parameter dq, however, is an extensive quantity, while the electrical potential (joules per coulomb) is the intensive quantity. For this reason, it is convenient to do yet another Legendre transform involving this term, so that the electrical potential becomes the differential variable:

$$\tilde{G} = G - \psi q \tag{4.71}$$

Taking the derivative gives

$$d\tilde{G} = dG + \psi \, dq + q \, d\psi \tag{4.72}$$

If the concentration gradient is included in the free energy expression, this equation reduces to the form

$$d\tilde{G} = -S \, dT + V \, dP + nRT(\ln c) + q \, d\psi \tag{4.73}$$

To reduce this equation to the chemical potential, 1 mole of charge is required. For an ion with charge z, there are zF coulombs per mole of ions where F is Faraday's constant (96,500 C/mole). For constant temperature and pressure, the new chemical potential is

$$d\tilde{\mu} = RT \, d(\ln c) + zF \, d\psi \tag{4.74}$$

This new potential is called the electrochemical potential and can be used to establish a balance between chemical and electrical driving potentials at equilibrium.

6. ELECTRICAL POTENTIALS ACROSS MEMBRANES

Let us consider the conditions necessary to create a potential difference across the membrane. As noted previously, the membrane cannot be permeable to both anions and cations. The problem is simplified further by assuming a single electrolyte. If the membrane is permeable to cations, there is then a single cation that can establish equilibrium between the two phases. If a concentration gradient exists for this ion, it will flow through the membrane carrying its charge with it. If Side 1 has the higher salt concentration, a positive charge excess will develop on Side 2, while Side 1 will be left with an excess of residual negative charge, as illustrated in Fig. 4.3. If a positive electrical test charge were placed between the solutions, it would move away from the positive charge on Side 2 toward the negative charges on Side 1; thus, a potential gradient would be created by the separation of the charges.

If the concentration gradient across the membrane is maintained, will this cationic flow continue indefinitely? If this were the case, the system could never reach equilibrium. Each transported cation would serve to increase the transmembrane potential. However, when this electrical potential becomes sufficiently large, it will oppose the flow of additional positive charge and the system will reach a point of dynamic equilibrium. The equilibrium is analogous to that observed for osmotic pressure. In this case, the electrical potential, not the osmotic pressure, opposes the ion flow induced by the concentration gradient.

To determine the magnitude of the potential difference that can be created by a given concentration gradient, the electrochemical potentials of the two bathing solutions must be equated:

$$RT \, d(\ln c_1) + zF \, d\psi_1 = RT \, d(\ln c_2) + zF \, d\psi_2 \qquad (4.75)$$

This equation is integrated through a sequence of equilibrium states to its final equilibrium configuration. There is no electrical potential, $\psi_1 = \psi_2 = 0$,

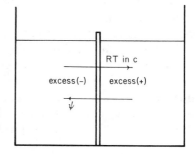

Fig. 4.3 Concentration and electrical potential, which balance at equilibrium for a cation-permeable membrane.

when the concentrations on both sides of the membrane are identical, so this is used as the equilibrium starting configuration. The final equilibrium concentrations are c_1 and c_2 as balanced by potentials of ψ_1 and ψ_2. The equation is

$$\int_c^{c_1} RT\, d(\ln c_1) + \int_0^{\psi_1} zF\, d\psi_1 = \int_c^{c_2} RT\, d(\ln c_2) + \int_0^{\psi_2} zF\, d\psi_2 \qquad (4.76)$$

$$RT[\ln(c_1/c)] + zF\psi_1 = RT[\ln(c_2/c)] + zF\psi_2 \qquad (4.77)$$

Eliminating common terms leads to

$$RT[\ln(c_1/c)] = ZF(\psi_2 - \psi_1) = zF\Delta\psi \qquad (4.78)$$

or

$$\Delta\psi = (RT/zF)[\ln(c_1/c_2)] \qquad (4.79)$$

The proportionality factor (RT/zF) has the units of volts. At 25°C, the factor RT/F is approximately 25 mV.

Although the selective ion permeability necessary to generate the potential in Eq. (4.79) might be expected to be a rare phenomenon, it is quite common in biological systems. For example, a nerve axon at rest will use its energetic machinery to maintain an internal concentration of K^+ that is 10 times higher than the external K^+ ion concentration. In addition, the two bathing solutions communicate via channels that are optimally selective for K^+ ion. The concentration gradient will lower the internal electrical potential relative to the external potential Because the external solution is generally assigned a potential of 0—that is, ground potential—the potential for the internal solution will be negative.

The magnitude of this potential difference is

$$\Delta\psi = (RT/F)[\ln(10/100)] = \psi_i - \psi_o$$
$$= (-25 \text{ mV})(2.3) = -57.5 \text{ mV} \qquad (4.80)$$

Although this potential is small (a dry cell has a potential of 1500 mV), it is sufficient to induce signal transmission when its magnitude is altered.

When multivalent ions are used in place of the univalent ion, a proportionately smaller potential difference is observed for the same concentration gradient. For example, if the permeant cation for Eq. (4.80) had been Ca^{2+}, Eq. (4.79) predicts that the observed potential will be only -23.8 mV. However, intuitively, it may appear that the potential should be twice as large. The Ca^{2+} experiences the same concentration gradient as the K^+ ion, so the same number of ions should move through the membrane while carrying twice the charge. Thus, the potential should be twice as great.

To resolve this difficulty, remember that the observed potential for a given concentration gradient represents a mutual balance point for the system. By rewriting the equation in terms of these balancing energies,

$$zF\Delta\psi = RT[\ln(c_1/c_2)] \tag{4.81}$$

it becomes apparent that the actual energy involved is the charge times the potential. The potential is halved, but two charges are carried down the gradient with each ion, so the actual work for the motion of one ion is the same.

7. THE NERVE ACTION POTENTIAL

When a membrane is selectively permeable to an ion and a concentration gradient of the ion appears across the membrane, an electrical potential is generated. For example, the concentration gradient of K^+ ions across a nerve axon membrane dominates production of an electrical potential of -60 mV. Interestingly, when the nerve is stimulated, this electrical potential will change with time, suggesting a departure from this equilibrium balance condition.

In the early studies of excitable membrane, no absolute comparison of the internal and external potentials was possible. Early workers postulated a mechanism in which the membrane became permeable to Cl^- when stimulated. Because a system permeable to both anions and cations will seek the condition of electroneutrality, the excitation phenomenon was simply the cancellation of the electrochemical potential via Cl^- ion flow.

With the discovery of the squid giant axon in the 1930s, electrodes could be placed both inside and outside the cell. The internal potential relative to the external ground reference was -60 mV before stimulation. When the nerve was stimulated by applying a short current pulse, the potential did indeed move toward zero potential but overshot this mark until it reached a transient potential maximum of $+50$ mV. The polarity of the transmembrane potential actually reversed during the excitation phenomenon. From this point, the system relaxed back toward negative potentials until stabilizing again at -60 mV. The entire process was thus a single-shot oscillation.

The source of the potential overshoot during stimulation can be traced to a second, latent electrochemical potential for the nerve system. The system also maintains an Na^+ ion concentration gradient with excess ions in the external bathing solution. However, the resting nerve axon is only weakly permeable to the Na^+ ions, and they are unable to establish this potential gradient, which

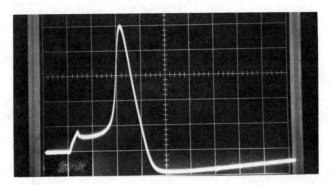

Fig. 4.4 Squid axon action potential. A stimulating current (lower left) produces a voltage excursion from -60 mV to $+50$ mV with subsequent return to -60 mV.

would have a magnitude of 110 mV. During stimulation, the nerve membrane becomes permeable to the Na^+ ion, and the resulting $+100$ mV is sufficient to cancel the -60 mV and produce an additional potential of $+50$ mV. As the Na^+ potential reaches its maximum, a second molecular process within the Na^+ channel inhibits the flow of Na^+ ions through the channel and the transmembrane potential will evolve to the potential dominated by the K^+ concentration gradient, -60 mV.

An action potential for squid giant axon is shown in Fig. 4.4. The transient Na^+ potential peak is reached in less than 2 ms. The mechanisms that produce this phenomenon will be explored in detail in Chapter 12.

8. THE RELATIVE ROLES OF IONS

The reversal of the transmembrane potential in the previous section was attributed to the 110 mV associated with the Na^+ concentration gradient. This raises an additional question for systems in which more than one ion may be permeable for a given membrane. Consider a membrane that is permeable to both Na^+ and K^+ ions. On Side 1, there are 1 M solutions of both NaCl and KCl, while on Side 2, both concentrations are 2 M, as shown in Fig. 4.5. There are now two distinct ways that we might use to calculate the electrochemical potential for this system. In the first case, assume that all the permeant cations will contribute to the concentration gradient. There are then permeant cations at 2 M on Side 1 and permeant cations at 4 M on Side 2. The electrical potential difference will then be

$$\Delta\psi = (RT/F)[\ln(4/2)] = -17.5 \text{ mV} \qquad (4.82)$$

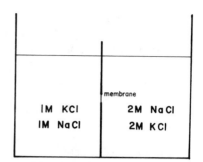

Fig. 4.5 Membrane permeable to both potassium and sodium ions.

In the second case, consider the potential that would be produced by each ion invididually. If KCl alone were added to the system, the potential difference would be

$$\Delta\psi = (RT/F) \ln(2/1) = -17.5 \text{ mV} \tag{4.83}$$

When the NaCl was then added to the solution, it should also produce

$$\Delta\psi = (RT/F)[\ln(2/1)] = -17.5 \text{ mV} \tag{4.84}$$

If each ion contributed independently of the other, the total potential difference across the membrane in this case would be -35 mV. Note that in the discussion of the nerve action potential, the Na^+ potential appeared to cancel the K^+ potential, as if they were generating their potentials independently.

The difference between the two approaches can be seen more clearly if the algebraic forms are compared. In Case 1, the total concentrations of permeant ions were considered. The potential difference is then

$$\Delta\psi = (RT/F)(\ln\{[c_1(K^+) + c_1(Na^+)]/[c_2(K^+) + c_2(Na^+)]\}) \tag{4.85}$$

For Case 2, the two potentials are added,

$$\Delta\psi = (RT/F)\{\ln[c_1(K^+)/c_2(K^+)] + \ln[c_1(Na^+)/c_2(Na^+)]\} \tag{4.86}$$

which can be reduced to

$$\Delta\psi = (RT/F)\{\ln[c_1(K^+)c_1(Na^+)/c_2(K^+)c_2(Na^+)]\} \tag{4.87}$$

In Case 1 we deal with a sum of concentrations, while in Case 2 a product of concentrations is required.

In Chapter 9, we will discover that the nature of the membrane does play a role in establishing the final equilibrium potential. The relative velocities at which the ions pass through the membrane will play a central role in establishing the observed potential. Thus, in the nerve axon example, the K^+ potential of -60 mV will dominate at equilibrium because the K^+ ion can pass through its channel much more effectively than the Na^+ ion can. Case 1 is then a more effective picture of the actual membrane equilibration process. However, in the actual solution, each of these concentrations must be weighted by a membrane-dependent parameter.

9. THE pH ELECTRODE

The pH, or glass, electrode constitutes a special case of a membrane system, where an H^+ concentration difference produces an electrochemical potential. Thus, the electrode can be used to determine H^+ ion concentrations in solution. The potential difference is

$$\Delta\psi = (RT/F)\{\ln[c(H^+)/c_{ref}]\} \tag{4.88}$$

The reference concencentration, c_{ref}, would normally be the concentration of the solution in the electrode interior. However, this would require careful preparation of this solution and replacement of the solution for different pH ranges. It is more convenient to measure two potentials; the first uses known external concentration, which is used as an external solution reference; the second unknown concentration is measured by observing the change in potential difference:

$$\Delta\psi = (RT/F)\{\ln[c(H^+)/c_{int}]\} - (RT/F)\{\ln(c_{ref, ext}/c_{int})\}$$
$$= (RT/F)\{\ln[c(H^+)/c_{ref, ext}]\} \tag{4.89}$$

In the pH meter, this output voltage is amplified and displayed on a meter calibrated in pH units.

Although the pH electrode provides a straightforward application of the electrochemical potential, it is also an excellent model system from which to pose additional questions about this potential. For example, we know that the transelectrode potential will be determined by the concentration gradient across this membrane, but we do not know exactly where this potential will be dissipated as a test charge moves through the electrode. For example, if the transmembrane potential is 100 mV, this entire 100 mV might be dissipated in transferring the proton from the external solution to a point just inside the electrode surface. From that point, the test charge would dissipate no further energy as it moved to the opposite solution. If we examine the potentials in each solution, the 100 mV potential drop would be observed; there is no way to know that this drop took place at the external surface.

This raises a second important question. If the proton generates the 100 mV drop by passing through the membrane surface, why should it move any farther? The electrochemical potential can be defined between any two phases with different ion concentrations. If the external solution is one phase, the second phase could be selected as the region just inside the membrane surface, and the entire potential difference will be concentrated at this interface. Thermodynamics requires no information on the size or physical properties of the phase as long as it can maintain a concentration of the ions. The problem could be resolved by placing one electrode just inside the surface and measuring the potential difference between this electrode and the external solution. However, this surface phase has molecular dimensions and it would be most difficult to position a conventional electrode properly.

In order to decide where the actual potential drop through the membrane takes place, consider the following possibilities:

(1) The ion is adsorbed at the membrane surface and creates a potential difference at this surface interface.

(2) The ion moves into, although not through, the membrane. Charge sites within the membrane will determine how much ion moves into this region relative to the concentration of ion in solution.

(3) The ion passes through the membrane to establish the potential difference.

All three "movements" could produce a change in the membrane potential difference, and, in fact, other regions of potential loss may be possible, such as the inner surface. However, only these three possibilities will be considered here. The total potential difference across the membrane would then be

$$\psi_t = \psi_{other} + \psi_{surf} + \psi_2 + \psi_3 \qquad (4.90)$$

where ψ_2 and ψ_3 are the potential drops associated with Processes 2 and 3 above, and ψ_{other} might include the inner surface potential drop, as noted, effects at the electrodes in the two bathing solutions, and any "asymmetry" potentials generated by differences in the glass structure at the inside and outside surfaces. The potential drops are illustrated in Fig. 4.6.

The bulk of the observed potential drop takes place inside the membrane, that is, ψ_2. The glass electrode functions as an ion exchange membrane. In such a system, anionic sites within the membrane play the role that the impermeant ion A^- played in our earlier study of the Donnan equilibrium. The sites are an intrinsic part of the glass membrane matrix and are therefore immobile. However, they can still act to control the relative amounts of ion between the external solution and the membrane, and these relative amounts then determine the electrical potential observed.

The protons do not pass through the membrane. This raises an additional question. In order to measure any potential, some current must flow through the membrane and the measuring device. Although the pH is determined by the relative concentrations of H^+ ions, currents through the membrane are carried by Na^+ ions.

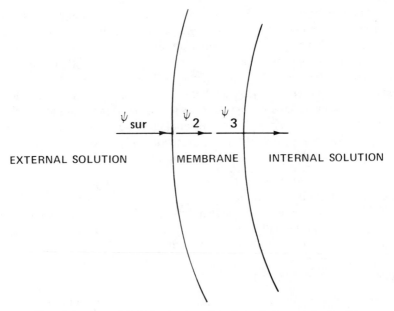

Fig. 4.6   Potential dissipation in a glass electrode (see text for details).

10. ULTRACENTRIFUGATION

In Chapter 1, we considered ultracentrifugation as a technique to determine the surface tension of intact cells. The gravitational energy of the centrifuge was used to counterbalance the membrane surface tension. The cell stretched and split in two along the direction of the applied gravitational field. Although this behavior can be described quantitatively using the energies for surface tension and gravitation, it is more convenient to use the cell forces, as the nature of the surface tension is more apparent in this format.

Consider a spherical egg that contains both an oil droplet (low density) and a nucleus (high density). If the cells are centrifuged, the two droplets will move at different rates and the cell will begin to stretch along the centrifuge axis. When the length of the cylinder thus formed exceeds the cylindrical circumference, the cell will split into two parts. Because the cell ruptures everywhere on the circumference, the force due to surface tension at the moment of rupture will be

$$\gamma(2\pi r) \tag{4.91}$$

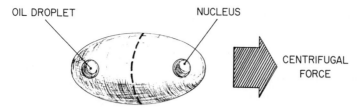

Fig. 4.7 Centrifuged cell at the instant of rupture.

This force must be balanced by the force produced by the centrifuge. This knowledge permits calculation of the surface tension for the membrane (Fig. 4.7).

In order to determine the force generated by the centrifuge, we consider the energy with which the particles are driven from the center. This is simply

$$U = mv^2/2 \qquad (4.92)$$

where v is the velocity of the particle. This equation can be reexpressed in terms of the angular velocity of the centrifuge,

$$v = \omega r \qquad (4.93)$$

where ω (radians/s) is the angular velocity and r is the distance from the centrifuge axis. The kinetic energy is

$$U = m\omega^2 r^2/2 \qquad (4.94)$$

The centrifugal force is determined from this equation by differentiating with respect to r,

$$dU/dr = m\omega^2 r = mv^2/r \qquad (4.95)$$

This force can now be balanced against the surface tension that resists stretching. However, we must include the effects of different masses for the two droplets in the cell, because this difference facilitates the stretching. Because the centrifuge creates a large gravitational field (the centrifugal force), the heavier droplet must move ahead of the medium while the lighter droplet will fall behind. If the droplets had a density equivalent to the medium, they could not change relative to the medium. The relative motion must then be related to the mass of a given volume of droplet less the mass of an equivalent volume of the medium. If the density of the nucleus is ρ_H and its volume is V_H while the density of the medium is ρ_M, the mass for use in the centrifuge equation is

$$m_H = V_H(\rho_H - \rho_M) \qquad (4.96)$$

If the density of lighter droplet is ρ_L and its volume is V_L, then its corrected mass is

$$m'_L = V_L(\rho_L - \rho_M) \tag{4.97}$$

The stretching force is proportional to the difference between these two masses,

$$F_s = (m'_H - m'_L)\omega^2 r \tag{4.98}$$

$$F_s = [V_H(\rho_H - \rho_M) - V_L(\rho_L - \rho_M)]\omega^2 r \tag{4.99}$$

This stretching force must equal the opposing surface tension force, so that

$$\gamma(2\pi r) = [V_H(\rho_H - \rho_M) - V_L(\rho_L - \rho_M)]\omega^2 r,$$

$$\gamma = [V_H(\rho_H - \rho_M) - V_L(\rho_L - \rho_M)]\omega^2 \tag{4.100}$$

The angular velocity of the centrifuge must be increased until the cell splits, and this velocity then leads directly to the surface tension of the cell.

PROBLEMS

1. Compare the osmotic pressure calculated using the formula $\Pi = cRT$ with the result obtained making no approximations for the chemical potential $RT(\ln X)$ for the case where $c_1 = 0\ M$ and $c_2 = 0.5\ M$. Assume an activity coefficient of 1.

2. The osmotic pressure generated between two solutions of concentrations $0.25\ M$ and $0.5\ M$, respectively, is counterbalanced with a mercury column. Determine the height of the column necessary to maintain this balance.

3. If the concentrations of solutions internal and external to a cell are different, an osmotic pressure flow can result. If the internal solution is more concentrated, develop an equilibrium relationship between this pressure difference and the surface tension of the cell membrane.

4. Baths 1 and 2 of a membrane permeable to K^+ and Cl^- have concentrations of $0.1\ M$ KCl. A $0.1\ M$ solution of KA is added to Side 1, while a $0.1\ M$ solution of K_2B is added to Side 2. Both A^- and B^{2-} are membrane-impermeable. Determine the final concentrations on each side of the membrane.

5. Bath 1 of a membrane contains a concentration c_1 of NaCl, while Bath 2 contains a concentration c_2 of NaCl and c_A of the sodium salt of a large anion with z charges. Develop the appropriate equations for equilibrium.

6. The relative concentrations of cation c_i^+ are often expressed as a Donnan ratio, $r = c_2^+/c_1^+$. Show that this ratio is also $r = c_1^-/c_2^-$.

7. As c_1 and c_2 approach each other, the logarithmic term in the expression for electrochemical potential can be expressed in terms of the concentration difference. Derive this relationship.

8. Determine the acceleration experienced by a particle at a distance of 10 cm from the axis of a centrifuge rotating at 50,000 rpm. Show the dimensions.

Chapter 5

Ion Transport

1. THE ELECTRODE INTERFACE

In Chapter 4 we demonstrated that ion concentration gradients could be used in conjunction with ion-selective membranes to produce electrical potential differences. In order to measure such potential differences, some contact with the external world—that is, a measuring device—is necessary. Electrodes establish the link between the electrochemical processes in the system and the flow of electrons in some external measuring circuit.

Consider the situation shown in Fig. 5.1 in which two platinum plates are introduced into the solutions bathing a membrane that supports an electrical potential difference of V volts. The right hand solution is selected as the reference potential, $V = 0$. When the two electrodes are connected externally as shown, electrons will flow from the electrode in Solution 2 to the electrode in Solution 1. However, current (charge flow) must be continuous throughout the entire loop. Because the electrons do not enter the solution, we must make a transition from electrons as charge carriers to the ions of solution as charge carriers.

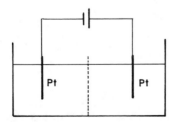

Fig. 5.1 Current in a simple electrolytic cell with two platinum electrodes.

91

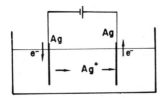

Fig. 5.2 Charge transfer at a silver electrode.

The transfer between electrode and solution occurs easily in situations when ions of the electrode material are present in the solution. Consider a silver metal electrode dipping into a solution of $Ag^+NO_3^-$ (Fig. 5.2). When the excess electrons reach the electrode, Ag^+ ions in solution may approach the surface, where they accept the electron and become Ag atoms on the electrode surface. The actual transfer process may be complicated, but the net effect is a silver ion flow to the surface that balances the electron flow. As each positive ion in solution is reduced at the surface, it leaves a residual negative ion in solution. If N electrons are neutralized at the surface, N excess negative ions will appear in solution. In the interfacial region, the negative electron flow is converted into a negative ion flow.

The excess negative ions cannot accumulate in solution. The continuity of current in the circuit requires the transfer of negative charges from the solution to the electrode for Solution 2. If this electrode were also silver, silver from the electrode would ionize, producing the required electrons in the electrical circuit and silver ions in the solution. The Ag^+ produced at this electrode would be equivalent to the Ag^+ lost at the opposite electrode, and the total solution would remain electroneutral.

The ions produced at the electrodes need not be metallic. Chlorine gas can be bubbled over an inert platinum electrode at some constant pressure, as shown in Fig. 5.3. If Cl^- ions are present in solution, transfer of current in

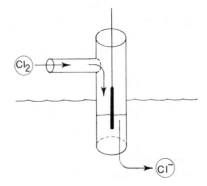

Fig. 5.3 The chlorine gas–chloride electrode.

either direction is possible. Each chlorine atom might accept an electron and enter solution as Cl^- or the chloride ion might surrender its electron to the platinum electrode, to become a chlorine atom. Hydrogen gas and ionic H^+ in solution will produce a hydrogen electrode, which is used as a standard reference electrode.

The chloride electrode presents some interesting experimental problems. It requires a source of chlorine gas that can dissolve in solution. The dissolved gas can affect the system under study. Silver ion "poisons" a number of biological phenomena observed in membrane systems. These difficulties detract from the advantages of these materials. For example, Cl^- ion is the dominant anion in most membrane systems because of the dominance of this ion in saline solutions. Silver metal is an excellent conductor, which makes it an excellent electrode material.

We can gather all the advantages of the Ag^+ and Cl^- systems by combining them into a single electrode. The electrodes discussed thus far are electrodes of the first kind. The electrode material and the ions produced in solution represent the same chemical species, such as H_2 and H^+, or Ag and Ag^+. Consider the electrode arrangement of Fig. 5.4. A layer of insoluble AgCl salt is plated on the silver surface of the electrode. This combined Ag–AgCl electrode is called an electrode of the second kind. The only chemical species that moves in and out of solution is the Cl^- ion, our ubiquitous solution ion.

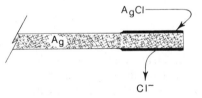

Fig. 5.4 The silver–silver chloride (Ag–AgCl) electrode, a reversible chloride electrode.

To illustrate the operation of the electrode, consider its operation when electrons appear in the silver electrode. These electrons reduce Ag^+ ion in the AgCl interfacial region, leaving a residual Cl^- ion. This ion then drifts into solution to continue the flow of negative charge with an acceptable ion. If electrons are removed from the electrode, they produce residual Ag^+ ions at the electrode surface. Although these Ag^+ ions might then be expected to drift into solution, they cannot do so because the AgCl salt is extremely insoluble. A Cl^- ion from solution will precipitate Ag^+ on the surface as AgCl. The net effect is the loss of Cl^- from the solution. Neither Cl_2 gas nor Ag^+ ion appears in the solutions.

2. ELECTRODES

The Ag–AgCl electrode is essentially reversible with respect to Cl⁻ ion. Other insoluble chloride salts can also be used to make reversible Cl⁻ electrodes. The most common of these electrodes is the calomel electrode, which uses mercury and insoluble calomel (Hg_2Cl_2) to form the reversible Cl⁻ electrode. We have already seen that a potential applied to such an electrode will result in the motion of Cl⁻ ion. However, the calomel electrode has the optimal ability to facilitate this transfer. In other words, for any current applied to the electrode, the calomel electrode will always tend to release or absorb the proper number of chloride ions. To an outside observer who could only observe charge, negative charge would appear to flow easily across the interface—that is, the interface would appear to have zero resistance to the flow of charge. Such an interface is called a nonpolarizable interface and can be described by the electrical equivalent circuit shown in Fig. 5.5. The resistance across the interface approaches zero, while the interfacial capacitance reflects the charge separation of ions in solution and charge on the interface. This capacitance is discussed in more detail in Chapter 8.

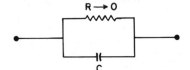

Fig. 5.5   Equivalent circuit for an ideal nonpolarizable electrode.

The ideal nonpolarizable electrode would have zero resistance. Different electrodes will exhibit varying levels of resistance. This suggests that we might also have electrodes with extremely high resistance. Such an electrode is called an ideal polarizable electrode when this resistance rises to infinity, as shown by the equivalent circuit of Fig. 5.6. For such an electrode, it becomes impossible to transfer the charge from the electrode to the solution. The charge remains on the electrode and attracts an equal and opposite charge to solution near the electrode.

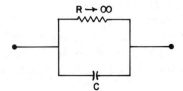

Fig. 5.6   Equivalent circuit for an ideal polarizable electrode.

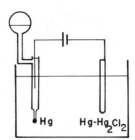

Fig. 5.7 An electrochemical cell with mercury (polarizable) and calomel (non-polarizable) electrodes.

The electrode normally selected as the best experimental approximation to the ideal polarizable electrode is the mercury (not calomel) electrode. A mercury drop is formed at the end of a fine capillary. Such electrodes have the advantage that they can be arranged to permit a steady flow of mercury drops. As one drop falls to the bottom of the container, a new drop with a clear surface is formed. The mercury metal is very unreactive, so it is difficult to get either Hg_2^{2+} or Hg^{2+} into solution.

Following our previous discussion of current flows via electrodes, the mercury electrode might appear to serve little purpose. When a potential is applied to the electrode, the capacitance will charge; when charging is completed, all current will stop and the electrode will assume the applied potential: it is polarized by the applied potential. However, consider the situation where an ideal polarizable and an ideal nonpolarizable electrode are both added to a common solution to form an electrochemical cell as shown in Fig. 5.7. When a potential V is applied between the two electrodes, the equivalent circuit of Fig. 5.8 results. Since the nonpolarizable electrode has zero resistance, the full potential must drop across the polarizable electrode. For this ideal system, we know the potential drops at each of the electrodes produced by the applied potential difference. For nonideal cases, potential drops will occur for each of interfaces.

The polarizable electrode will always assume the voltage applied to it, while the low resistance of the nonpolarizable electrode indicates that it will always provide a low-resistance current path in the system. For systems that require large current flows, the nonpolarizable electrodes are then the optimal choice.

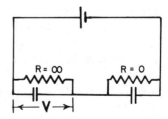

Fig. 5.8 Equivalent circuit for the mercury–calomel system, illustrating the full potential drop across the polarizable electrode.

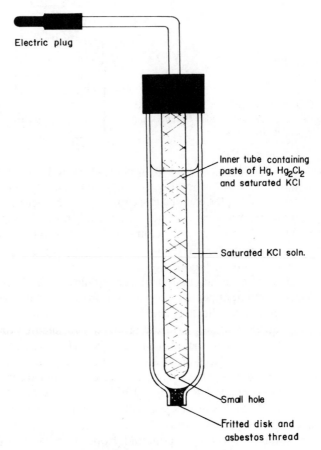

Fig. 5.9   The calomel electrode.

The calomel electrode is fairly complicated, as shown in Fig. 5.9. This makes it quite difficult to miniaturize for use in cellular systems. The Ag–AgCl electrode, on the other hand, can be formed using a length of silver wire. The wire is placed in a solution of HCl or KCl and connected to the positive terminal of an external potential source, as shown in Fig. 5.10. A layer of AgCl will grow on the surface during this plating process. If the electrode resistance is measured during the plating process, it will decrease as the electrode becomes progressively more nonpolarizable. The practical problem here lies in finding the optimal plating regimen to produce the lowest possible resistance.

An alternative electrode that can also be developed around a length of wire or a plate is the plated platinum electrode. A platinum electrode is electrolyzed

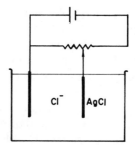

Fig. 5.10   Preparation of an Ag–AgCl electrode.

in a solution of chloroplatinic acid to produce a nonpolarizable electrode that is reversible to chloride. Since even trace amounts of silver can often harm biological preparations, the plated platinum electrode is often preferred if the electrode is to be used in a cell interior. For example, plated platinum electrodes are used in voltage clamp systems (Chapter 6) as the current-carrying electrode.

Both the Ag–AgCl and plated platinum electrodes have advantages that make them useful. The Ag–AgCl electrode has a higher resistance, but the potential drop across the electrode tends to remain stable with time. The plated platinum electrode has a lower resistance. For this reason, Cole developed a hydrid electrode. An Ag–AgCl electrode was platinized to produce an electrode with both good potential stability and low resistance.

3. INTERFACIAL POTENTIALS

All electrodes except the ideal nonpolarizable electrode will have a finite resistance, so that the potential drop in the electrode region is finite. We must now consider the physical microscopic processes at the electrode that give rise to this potential drop. To do so, we must consider all the potential differences that result when a potential is applied between two electrodes. Each of the electrodes will have a potential drop. In addition, other potentials may appear at any junction between two disimilar metals. For example, if a copper wire from the external potential is connected to a silver electrode, a potential difference, the junction potential, will appear at this connection. If the two electrodes are made of different metals, the two junctions will produce a net potential that must be subtracted from the total potential difference to determine the potential difference across the electrodes and solution. If both electrodes are made of the same materials, the two junction potentials will cancel and a correction is not necessary.

To simplify the discussion, consider an electrode matched with an ideal nonpolarizable electrode so that the full potential drop lies across the electrode of interest. This is an ideal situation, because the ideal nonpolarizable electrode is a limiting construct. In an actual experiment with two real electrodes, we cannot know the potential differences across the individual electrodes. These could only be determined by inserting microscopic electrodes just inside the electrode and in solution just outside the electrode; to be successful, the potential drop across these additional electrodes would have to be known exactly. If such electrodes existed, they could simply be used in place of the electrode under study.

The equivalent circuit of the interface indicated both a capacitive and a resistive component. If the interface serves to separate a certain amount of charge, the charge separation will produce a potential difference. Thus, at least some of the observed potential drop must be manifested in this charge separation. To define the potential due to the stored charge, a test charge is moved from infinity, where its potential is zero, to the region of stored charge in solution. The work required for this process is described by the potential ψ_S. If the same test charge is then moved from infinity to the electrode surface, a different potential ψ_E is measured. The potential difference across the interface due to the stored charge is then

$$\psi = \psi_E - \psi_S \qquad (5.1)$$

as shown in Fig. 5.11. We have not defined the regions in the solution and the electrode to which we bring the test charge. As the charge approaches the surface of either phase, it will begin to induce image charges in the second phase. To avoid this complication, we arbitrarily choose our regions sufficiently far from the surface so that image charges are not important. The potential difference ψ defined by the separated charge regions is called the Volta potential.

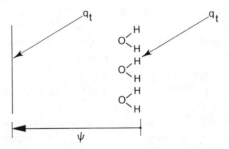

Fig. 5.11 Determination of the interfacial potential difference (the Volta potential) between the surface and the stored charge region. A test charge q moves from infinity to each surface and ψ is the difference between the potentials required to reach each surface.

The Volta potential is defined as the difference in potentials required to move a test charge to the solution region and the electrode region. Under such conditions, we ignore other energetic processes that might take place in the intermediate interfacial region. Although our separated charges may not enter this intermediate region, this does not preclude electrostatic effects here. Water molecules are present in this region, and, because of the electrode surface, they assume a preferred orientation. The potential of each dipole is proportional to $\cos \theta$ (Chapter 3), so that one side of each oriented dipole will display a positive potential while that of the opposite side is negative. A test charge passing this array of oriented dipoles will thus pass through a potential difference. A similar situation can arise within the electrode itself. For example, electrons may move to the surface creating a charge separation (dipole) between these electrons and their positive ions in the electrode. The potential difference generated by these two dipolar regions is called the dipole potential, χ.

For our model of a nonpolarizable electrode and our electrode of interest, we can now define the total potential drop across the electrode as the sum of the Volta and dipole potentials. The total potential drop is the Galvani potential difference ϕ,

$$\phi = \psi + \chi \tag{5.2}$$

Now, consider a system that has been corrected for junction potentials,

$$V' = V_{\text{ext}} - V_{i,\,\text{jp}} \tag{5.3}$$

This net potential difference must be equivalent to the sum of all potential drops in the system. If we chose two real electrodes, as shown in Fig. 5.12, this net potential difference must be equivalent to the Galvani potential drops for the two electrodes, so that

$$V' = \phi_{\text{E}_1,\,\text{S}_1} + \phi_{\text{E}_2,\,\text{S}_2} \tag{5.4}$$

Since there are always a minimum of two electrodes in the system, it is impossible to resolve the observed potential difference V' into the two Galvani potentials. To determine the individual ϕ terms, we require detailed information on the interfacial region. Such information is required to determine the dipole potential χ for each region.

Fig. 5.12 The Galvani potential as the net potential difference in a two-electrode system. E_1—electrode 1; S_1—solution 1.

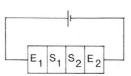

4. ION MOTIONS IN SOLUTION

For the reversible Ag–AgCl electrode, Cl$^-$ ions were either adsorbed as AgCl or released from the AgCl of the surface as Cl$^-$ ions. For either of these processes to occur, the ions had to move relative to the electrode in solution. These ionic motions obviously constitute the current in the solution phase, and, for this reason, the measurable parameters associated with such motions must be investigated.

If two nonpolarizable electrodes are placed into a solution and an external potential difference is applied, a current will flow through the solution. This current is obviously associated with ionic motions. However, to determine an experimental parameter for study, we determine the resistance of the solution using Ohm's law,

$$R = V/I \tag{5.5}$$

Alternatively, we can define a conductance G as

$$G = I/V \tag{5.6}$$

which is simply the inverse of the resistance,

$$G = 1/R \tag{5.7}$$

These parameters will depend on the particular ions in solution, their concentrations, and the size of the container in which the measurements are made. For this reason, a standard-size measuring unit is preferred.

Consider a block of metal that is connected to a potential difference at two opposite faces, as shown in Fig. 5.13. As the cross-sectional area of the block increases, the conductance G of the block should increase, because more electron pathways are available. The conductance is directly proportional to the cross-sectional area A. On the other hand, as the thickness L of the block increases, the conductance should fall, because more energy must be dissipated by an electron crossing the block. This inverse dependence and the area dependence suggest a proportionality relation of the form

$$G = \kappa A/L \tag{5.8}$$

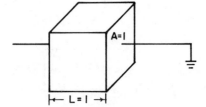

Fig. 5.13 Unit conducting block used to define the conductivity, Λ.

where κ, the conductivity, is the conductance for a block of metal of unit cross-sectional area and unit length. By using the conductivity, all measurements are normalized to a cube of unit cross section and unit length.

A conductivity can also be defined for a unit cube of solution. However, this conductivity will differ for different ion species and their concentrations. Thus, the conductivity is normalized still further by defining a conductivity per unit ion concentration. This normalized conductivity is called a molar conductivity,

$$\Lambda = \kappa/c \tag{5.9}$$

The molar concentration for such measurements is the total moles of salt in a unit cube of solution. In cgs units, the unit cube is $1\ \text{cm}^3$.

The use of a salt concentration for this operational definition leaves something to be desired, because molar conductivity must be determined for each salt. In addition, the molar conductivity varies with the concentration, so that the normalized conductivity will be different at different concentrations. This situation can be rectified by noting that a plot of Λ versus the square root of the concentration is linear, as shown in Fig. 5.14. For this reason, it is convenient to extrapolate to zero concentration and define a molar conductivity at infinite dilution, Λ_o. In the infinite dilution region, where interactions between ions are minimal, each ion carries charge independently and the molar conductivities can be partitioned into the conductivities of the individual ion species,

$$\Lambda_o(C^+A^-) = \lambda_o(C^+) + \lambda_o(A^-) \tag{5.10}$$

These individual ion molar conductances can be combined to give the total molar conductivity for a given salt. For example, the molar conductivity of $MgCl_2$ is

$$\Lambda_o(MgCl_2) = \lambda_o(Mg^{2+}) + 2\lambda_o(Cl^-) \tag{5.11}$$

One further normalization is possible. An equivalent conductivity is the molar conductivity per unit ionic charge,

$$\lambda'_o = \lambda_o/z \tag{5.12}$$

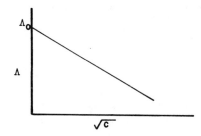

Fig. 5.14 Equivalent conductivity Λ versus $\sqrt{c}$; Λ_o is the equivalent conductivity at infinite dilution.

where z is the charge on the ion. All these definitions are related and provide some insight into the motions of the ions in solution.

The molar conductivities for K^+ and Cl^- are 74S m^2/mole and 76 S m^2/mole, respectively, in MKS units. (1 S = 1 Siemen = 1 ohm^{-1}). When the ions experience the common applied potential, they carry currents that are proportional to their conductivities. Because the Cl^- conductivity is slightly larger, it is expected to carry a larger fraction of the current. The fraction of current carried by the K^+ ion, t_+, is

$$t_+ = \lambda_o(K^+)/[\lambda_o(K^+) + \lambda_o(Cl^-)] = \lambda_o(K^+)/\Lambda_o(KCl) \qquad (5.13)$$

Here, t_+ is the transport number for K^+, while the transport number, t_-, for the Cl^- ion is

$$t_- = \lambda_o(Cl^-)/\Lambda_o(KCl) \qquad (5.14)$$

Using the molar conductivities for K^+ and Cl^- gives the transport numbers $t_+ = 74/(76 + 74) = 0.49$ and $t_- = 76/150 = 0.51$. The fractions of current carried by the K^+ and Cl^- are almost equal. For an NaCl solution, the transport numbers are $t_+ = 0.4$ and $t_- = 0.6$.

The molar conductivities ultimately describe the current that flows when a potential of 1 V is applied to a unit volume of unit concentration. The current per unit voltage is a measure of ion velocities in the resultant electric field. Because the ion will move at different velocities for different applied potentials, we define a mobility as the velocity per unit electric field.

$$u = v/E \qquad (5.15)$$

This mobility relates to the equivalent conductivity through the Faraday constant to give a relation of the form

$$\lambda_i = u_i F \qquad (5.16)$$

Equation (5.16) can be verified by checking units for both sides. The left side has the units

$$\lambda_i: \quad S\ m^2/mol \qquad (5.17)$$

while the right side of the equation has units

$$u_i F: \quad [(m/s)/(V/m)](C/mol) = [(C/s)/V](m^2/mol) \qquad (5.18)$$

$$= S\ m^2/mol \qquad (5.19)$$

When the ionic molar conductivities in solution are known, the mobilities for the ions can be calculated directly from them.

The equations for the different ionic parameters have been developed around 1:1 electrolytes. Slightly different definitions are often used when polyvalent ions are present. For example, mobilities are often defined in terms of unit charge. The velocity in a field E is then

$$v = zuE \tag{5.20}$$

where z is the total charge on the ion. The transport numbers for each ion must also be redefined when the solution contains multiple ions of varying ionic charge. If we define all mobilities u as the mobility per unit charge, then Eq. (5.20) defines the velocity of the ion. However, the current carried by the ion will also depend on its concentration in solution. Thus,

$$i_i \propto u_i |z_i| c_i \tag{5.21}$$

for the ith ion. The absolute value of z_i indicates that ions of either charge can produce the final net current.

The proper transport number for the ith ion is now

$$t_i = (u_i |z_i| c_i) / \sum (u_i |z_i| c_i) \tag{5.22}$$

Because molar conductivity and mobility satisfy the equation

$$\lambda_i = z_i u_i F \tag{5.23}$$

the result is that the equivalent conductivity,

$$\lambda_i' = \lambda_i / z \tag{5.24}$$

now becomes

$$\lambda_i' = F u_i \tag{5.25}$$

The mobility per unit charge will be used in the sections that follow.

5. CONCENTRATION CELLS

The chemical potential for an ion at some concentration c is defined

$$d\mu = RT \, d(\ln c) \tag{5.26}$$

while the electrochemical potential is

$$d\tilde{\mu} = RT \, d(\ln c) + zF \, d\psi \tag{5.27}$$

For equilibrium situations, Eq. (5.27) suggests that any change in the concentration will be accompanied by a change in ψ to maintain the system at equilibrium, that is, $d\tilde{\mu} = 0$. Thus, a potential change can be expressed in terms of the concentration change,

$$-d\psi = (RT/zF) \, d(\ln c) \tag{5.28}$$

for 1 mole of a specific ion. Further, 1 coulomb of charge will transfer 1 mole of the ion.

For a 1:1 electrolyte, the potentials associated with each ion must be included. This presents a problem, because the two ions must divide a coulomb of charge passed. The fraction of total charge carried by an ion is given by its transport number. The total potential change will now be the sum of the potential changes produced by each of the ions weighted by their respective transport numbers,

$$-d\psi = (RT/F)\left[\sum(t_i/z_i)\,d(\ln c_i)\right] \qquad (5.29)$$

The potentials produced in solution now depend both on the concentration gradients and the relative velocities of the respective ions.

For salt solutions of different concentration, a potential difference can appear if the solutions are coupled properly. Consider the situation of Fig. 5.15, where the two solutions are coupled by a pair of connected polarizable electrodes. Since $c_2 > c_1$, the net ion flow must move toward Cell 2. However, since both H^+ and Cl^- have identical concentrations in Cell 2, both ions will interact with the electrode and no preferred ion flow leading to a potential difference is possible. The situation is similar to that of two solutions separated by an ion-impermeable membrane. The potential is present but it cannot be tapped.

In order to obtain a useful potential from the HCl gradient, we must provide electrodes that permit separate paths for H^+ and Cl^-. In Fig. 5.16, non-polarizable electrodes reversible to either H^+ or Cl^- are placed in each cell and the two Cl^- electrodes are connected. The reversible electrodes function as ion-selective junctions. The larger chloride concentration of Cell 2 is spontaneously transferred to Cell 1 when a Cl^- ion is deposited as AgCl on the Ag–AgCl electrode of Cell 2 and a Cl^- ion is released from the Ag–AgCl electrode of Cell 1. Because the negative ions now flow to the hydrogen electrode of Cell 1, this electrode becomes the anode of the total cell.

The excess H^+ ions of the cell operate similarly at the reversible hydrogen electrode. The excess H^+ of Cell 2 will move toward the electrode in Cell 2,

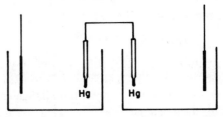

Fig. 5.15 Mercury electrodes linking two solutions of different HCl concentrations (c_1 and c_2).

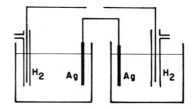

Fig. 5.16 An electrode system to extract the maximal electrochemical potential from two HCl solutions of different concentration.

the cathode. If the two hydrogen electrodes are connected, this will produce a spontaneous flow of positive current from Cell 2 to Cell 1 and the release of H^+ into Cell 1. With this arrangement, both potentials from both ions combine to produce a maximal net potential. The positive potential difference observed at the hydrogen electrode of Cell 2 is the sum of the two potentials,

$$\psi = (RT/F)\{\ln[c_{Cl}(1)/c_{Cl}(2)] + \ln[c_H(1)/c_H(2)]\} \qquad (5.30)$$

$$= (RT/F)\ln[c_{Cl}(1)c_H(1)/c_{Cl}(2)c_H(2)] \qquad (5.31)$$

At higher concentrations, when ionic interactions become important, the concentrations must be replaced by activities. In addition, because such interactions are mutual, it becomes impossible to separate the energetic contributions of each ion. A geometric mean activity,

$$a^2 = a_+ a_- \qquad (5.32)$$

or

$$c^2 = c_+ c_- \qquad (5.33)$$

is used in Eq. (5.31) to give

$$\psi = 2(RT/F)\ln[a(1)/a(2)] \qquad (5.34)$$

Because the concentrations of univalent anions and cations must be equal, the factor of 2 arises automatically when the concentrations are used.

$$\psi = 2(RT/F)\ln[c(1)/c(2)] \qquad (5.35)$$

Equation (5.35) represents the maximum potential possible for a given concentration gradient. We can derive this same result in an alternate way by driving a coulomb of charge through the system and determining the amount of HCl transferred. In this case, both concentrations are initally equal. One coulomb of charge moves from the H^+ electrode in Cell 1 and is transferred to the H^+ electrode in Cell 2. One mole of H^+ is released into Solution 1 while 1 mole of H^+ disappears from Solution 2, as shown in Fig. 5.16. The Ag–AgCl electrode of Solution 1 maintains the electroneutrality by releasing 1 mole of Cl^-, while 1 mole of Cl^- is precipitated in Solution 2. The net effect is the transfer of 1 mole of HCl. The potential necessary to accomplish this transfer must then relate to the final solution concentrations as Eq. (5.31).

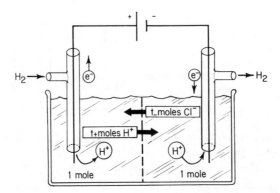

Fig. 5.17 A concentration cell with transport between the chambers of different concentration.

Consider the situation of Fig. 5.17, in which the two Ag–AgCl electrodes are replaced with a porous partition so that both H^+ and Cl^- can cross the internal boundary. As these ions have opposite polarities, we might expect this potential difference to be smaller, because some of the polarity used to generate the Cl^- potential is now cancelled by the concomitant flow of H^+ ion. Intuitively, we might expect the potential to be lowered by the fraction of charge carried by H^+ ion, that is,

$$\text{Potential} = 2(1 - t_+)(RT/F)\ln[c(1)/c(2)] \qquad (5.36)$$

$$= 2t_-(RT/F)\ln[c(1)/c(2)] \qquad (5.37)$$

We can demonstrate that this equation is indeed correct by considering what happens when a coulomb of charge is added to Solution 2 and removed from Solution 1. The process is illustrated in Fig. 5.17. One coulomb of charged H^+ ion is added to Solution 1 at the electrode. However, the H^+ in solution carries t_+ of the current from Solution 1 to Solution 2. The net gain of H^+ in Solution 1 is only $(1 - t_+) = t_-$ moles of H^+; similarly, Cl^- ion moves from Solution 2 to Solution 1 because of its negative sign, but only t_- moles of Cl^- are transferred. A total of t_- HCl appears in Solution 1 after passage of the charge. The resultant potential is then reduced by the factor of t_- to verify the result of Eq. (5.37).

These electrochemical cells with and without transference provide a technique for the measurement of transference numbers. The potentials are measured for both cells, and the ratio of the potentials is equal to the transference number

$$t_- = \psi_{wt}/\psi_{nt} \qquad (5.38)$$

The expression gives the transference number for the ion that has no reversible electrode connection.

6. THE DIFFUSION POTENTIAL

The potential between two solutions of different concentration was maximal when all electrodes were reversible to ions and decreased when the reversible electrodes for one of the ions were replaced by direct contact between the solutions. Although the lost portion of the potential was not present at the electrodes, conservation of energy requires that it must have been dissipated somewhere in the solution. The cations and anion crossed the boundary between the two solutions at different rates, because of their different mobilities. The resultant separation of charge can account for the missing potential.

To measure such a diffusion potential, we insert two polarizable electrodes into two solutions of different concentrations arranged to permit transfer of both ions. The cations and anions will both flow toward the lower concentration solution, but a greater fraction of one of the ions will cross the boundary because of the larger mobility of the ion. The potential will then be determined by Eq. (5.29), which is

$$-d\psi = (RT/F)[\sum(t_i/z_i)\,d(\ln c_i)] \tag{5.39}$$

For the univalent cation and anion, the equation is

$$-d\psi = (RT/F)[t_+\,d(\ln c_+) - t_-\,d(\ln c_-)] \tag{5.40}$$

In order to determine the potential, we must integrate across the region of solution mixing until we reach unmixed solution. Thus, the equation must be integrated from $c_H(1)$ and $c_{Cl}(1)$ to $c_H(2)$ and $c_{Cl}(2)$, respectively. For the univalent–univalent case, the concentrations and charges on both species are identical and the transport numbers can be expressed in terms of mobilities as

$$t_+ = u_+/(u_+ + u_-) \tag{5.41}$$

and

$$t_- = u_-/(u_+ + u_-) \tag{5.42}$$

Both t_+ and t_- are constants that can be removed from the integral, and the final result is

$$-\psi = (RT/F)(t_+\{\ln[c_H(2)/c_H(1)]\} - t_-\{\ln[c_{Cl}(2)/c_{Cl}(1)]\}) \tag{5.43}$$

$$= (RT/F)(t_+ - t_-)\{\ln[c(2)/c(1)]\} \tag{5.44}$$

If Eqs. (5.41) and (5.42) are inserted for t_i, the expression is

$$-\psi = (RT/F)[(u_+ - u_-)/(u_+ + u_-)]\{\ln[c(2)/c(1)]\} \tag{5.45}$$

The diffusion potential is proportional to the difference in cation and anion mobilities. If the cation moves faster than the anion, for example, positive charge will move along the concentration gradient at a faster rate, producing a separation of positive and negative charge. Such a charge separation gives rise to a potential. However, the separation cannot continue indefinitely. As the potential increases, the increased numbers of unbalanced positive and negative charges will exert an increasing attraction. A stable potential results when the differential mobility produced by the concentration gradient is just balanced by the potential produced by the charge separation. The result is the steady diffusion potential observed.

The diffusion potential results from the differential mobilities of ions within a phase. However, it can be eliminated by selecting anions and cations that have similar mobilities, because the difference will approach zero. We showed previously that K^+ and Cl^- each carried about 50% of the current because their molar conductivities were almost equal. Because these conductivities are directly proportional to the ionic mobilities, a KCl solution will generate a very small diffusion potential. For this reason, it is often used as a salt bridge between two solutions. A tube filled with KCl connects the two solutions of interest. The KCl is often mixed with 2% agar to produce a gel within the tube. The high conductivity of the KCl solution is maintained, but the gel remains in the tube during handling.

A salt bridge may be used to connect a biological cell and the Ag–AgCl electrode. A capillary with a fine tip is filled with a KCl solution and is then

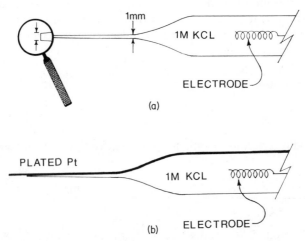

Fig. 5.18 (a) Ag–AgCl electrode with KCl salt bridge. (b) A piggyback electrode containing the Ag–AgCl salt bridge of (a) and a plated platinum electrode to carry current.

connected to a larger KCl reservoir, which contains the Ag–AgCl electrode [shown in Fig. 5.18(a)]. The fine capillary tip is then used to penetrate to the cell interior. Because the capillary tip must be only a few micrometers in diameter for penetration of the cell without damage, such a salt bridge–electrode arrangement constitutes a high-impedance system with resistances in the megohm range. An alternative system, which is used for axial entry into cylindrical cells such as nerve axons, is shown in Fig. 5.18(b). For voltage clamp experiments, a current-carrying electrode is attached to a capillary salt bridge for voltage sensing. This piggyback electrode is then inserted axially into the axon. Because the salt bridge capillary does not have to be drawn to a fine tip, this electrode system is a low-impedance system.

7. THE HENDERSON EQUATION

To determine the diffusion potential for univalent ions, we integrated over the region where the two concentrations were mixing. Our task was simplified considerably by the fact that the transport numbers were constant throughout the region. If many ionic species with a variety of charges are present the analysis becomes more involved, because the transport numbers depend on the concentrations of the different ionic species. In such a case, the transport numbers must remain within the integral. In addition, these transport numbers will then vary across the total mixing region of the integrals.

Henderson postulated a linear variation of concentration across the mixing region, as shown in Fig. 5.19. The region is described by an arbitrary distance parameter x, which varies from 0 to 1: that is, x describes a fractional distance across the mixing region, rather than some absolute distance. The concentration of the ith ion now varies linearly via the equation

$$c_i = [c_i(1)](1 - x) + [c_i(2)]x \qquad (5.46)$$

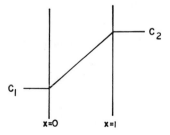

Fig. 5.19 The linear variation of concentration between two fixed concentration regions $0 \leq x \leq 1$.

The limiting cases can be checked by substituting $x = 0$ or $x = 1$. Because $d(\ln c)$ is our variable of interest, Eq. (5.46) can be used to define this variation as

$$d(\ln c_i) = dc_i/c_i = -[c_i(1) - c_i(2)] dx/c_i \tag{5.47}$$

for each of the i ions. Because the concentrations vary with x, the transport numbers must also vary,

$$t_i = c_i z_i u_i / (\textstyle\sum c_i z_i u_i) \tag{5.48}$$

$$= [c_i(x)] z_i u_i / (\textstyle\sum z_i u_i [c_i(1)] + x \{\sum z_i u_i [c_i(2) - c_i(1)]\}) \tag{5.49}$$

These expressions must now be substituted into the terms

$$(t_i/z_i) \, d(\ln c_i) \tag{5.50}$$

and integrated. Substituting gives

$$\frac{[c_i(x)] u_i z_i [c_i(2) - c_i(1)] \, dx}{[c_i(x)] z_i \{\sum z_i u_i c_i(1) + x \sum z_i u_i [c_i(2) - c_1(1)]\}} \tag{5.51}$$

The factor $[c_i(x)] z_i$ cancels. To facilitate integration of the expression, the following definitions are introduced:

$$a = u_i [c_i(2) - c_i(1)] \tag{5.52}$$

$$b = \textstyle\sum z_i u_i c_i(1) \tag{5.53}$$

$$d = \textstyle\sum z_i u_i [c_i(2) - c_i(1)] \tag{5.54}$$

so that Eq. (5.51) reduces to the form

$$(t_i/z_i) \, d(\ln c_i) = a \, dx/(b + dx) \tag{5.55}$$

All i terms in the sum must now be integrated for x from 0 to 1,

$$\int_0^1 [a/(b + dx)] \, dx = (a/d) [\ln(b + dx)]\big|_0^1 \tag{5.56}$$

$$= (a/d) \{\ln[(b + d)/b]\} \tag{5.57}$$

The expression for the total transmembrane potential is then determined by summing terms of the form of Eq. (5.57) over all ions of the system,

$$\psi = -(RT/F) \textstyle\sum ((a/d) \{\ln[(b + d)/b]\})$$

$$= -(RT/F) \frac{\{[\sum c_i(2) - c_i(1)] u_i\}}{\sum u_i z_i [c_i(2) - c_i(1)]} \left(\ln\left\{ 1 + \frac{\sum [c_i(2) - c_i(1)] z_i u_i}{\sum [c_i(1)] z_i u_i} \right\} \right) \tag{5.58}$$

Although this final expression appears quite complicated, a closer examination indicates that all the terms are sums of two types. The sums may simply add concentrations, or they may sum concentrations weighted by their mobilities. The equation itself can be simplified still further by changing the factor 1 in the logarithmic term to a ratio of the term

$$\sum [c_i(1)]z_i u_i$$

so that the term in $c_i(1)$ is eliminated from the numerator to give

$$\psi = (RT/F) \frac{\{\sum [c_i(2) - c_i(1)]u_i\}}{\{\sum [c_i(2) - c_i(1)]z_i u_i\}} \ln\{\sum [c_i(2)]u_i z_i / \sum [c_i(1)]u_i z_i\} \quad (5.59)$$

Now the various cation and anion concentration sums are collected via the following definitions

$$C_1 = \sum \{[c_+(1)]u_+\}_1 \qquad A_1 = \sum \{[c_-(1)]u_-\}_1$$
$$P_1 = \sum \{[c_+(1)]u_+ z_+\}_1 \qquad M_1 = \sum \{[c_-(1)]u_- z_-\}_1 \qquad (5.60)$$

Note that the factor z_- can introduce a sign change. When these factors and the corresponding factors for Side 2 are substituted into the equation, it becomes

$$\psi = -(RT/F)\{[(C_2 - A_2) - (C_1 - A_1)]/[(P_2 + M_2) - (P_1 + M_1)]\}$$
$$\times \{\ln[(P_2 + M_2)/(P_1 + M_1)]\} \qquad (5.61)$$

In this expression, the anion terms A_i appear with a negative sign, which reflects the fact that anion mobilities will be opposite to the cation mobilities. In the P and M terms, which include the ionic charge, the negative motion of the ions and the negative charge on the ion will produce a net positive. Thus, the terms with P and M are sums.

Henderson's equation is more complicated than the simple diffusion potential equation, which was derived assuming that the mobilities were constant. However, it is worth noting the differences for this equation if we consider only the case for a single univalent anion and cation. Under these circumstances, the appropriate factors become

$$\Delta\psi = -(RT/F) \frac{[c(2)](u_+ - u_-) - [c(1)](u_+ - u_-)}{[c(2)](u_+ + u_-) - [c(1)](u_+ + u_-)} \ln \frac{[c(2)](u_+ + u_-)}{[c(1)](u_+ + u_-)}$$

$$= -(RT/F)[(u_+ - u_-)/(u_+ + u_-)]\{\ln[c(2)/c(1)]\} \qquad (5.62)$$

because the transport numbers do not depend on concentration. In this case, the Henderson equation reduces to the previous equation for diffusion potential.

A second limiting case is worth considering because it appears frequently in biological membrane systems. In this case, there are two cations (Na^+ and K^-) and a common anion (Cl^-). The relevant parameters are derived when there are equal concentrations c of NaCl and KCl on opposite sides of the interfacial region. The relevant parameters are then

$$C_1 = cu_{Na} \qquad\qquad C_2 = cu_K$$

$$A_1 = cu_{Cl} \qquad\qquad A_2 = cu_{Cl}$$

$$P_1 = cu_{Na} = C_1 \qquad P_2 = cu_K = C_2$$

$$M_1 = cu_{Cl} = A_1 \qquad M_2 = cu_{Cl} = A_2$$

and the potential difference is

$$\psi = -(RT/F)\frac{(cu_K - cu_{Cl} - cu_{Na} + cu_{Cl})}{(cu_K + cu_{Cl} - cu_{Na} - cu_{Cl})}\ln\frac{c(u_{Na} + u_{Cl})}{c(u_K + u_{Cl})}$$

$$= -(RT/F)\{\ln[(u_{Na} + u_{Cl})/(u_K + u_{Cl})]\} \qquad (5.63)$$

Because the mobilities provide a relative measure of the current that can be carried by the ions, the logarithmic ratio is simply related to the relative conductivities of each solution, that is,

$$\Lambda = \lambda_{Na} + \lambda_{Cl} \propto u_{Na} + u_{Cl} \qquad\qquad (5.64)$$

and the expression reduces to

$$\psi = -(RT/F)[\ln(\Lambda_{NaCl}/\Lambda_{KCl})] \qquad\qquad (5.65)$$

While these limiting expressions are important, the basic form of the Henderson equation also provides some insight into the relative roles of different ions in establishing the diffusion potential. In order to see this, it is necessary to examine the summations that appear in the logarithmic portion of the expression. The terms have the form

$$\sum c_i u_i z_i \qquad\qquad (5.66)$$

Unlike some earlier cases that included only the concentration, this expression has concentrations that are weighted by their mobilities. In other words, the equation suggests that the ions that move more rapidly through the solution will play the greater role in determining the net membrane potential. These mobilities thus play the roles of activity coefficients in this equation. A similar result will occur when the Nernst–Planck equation is analyzed.

Although the Henderson equation postulates a linear change in ionic concentration across the region of interest, it says nothing about the electric field distribution in this region. However, the linear distribution of ionic charge implies some regularity in the electric field. This will be explored in Chapter 9.

PROBLEMS

1. The equivalent conductivity of aqueous NaCl is $\Lambda_o = 128.1 \, \text{S} \cdot \text{cm}^2/\text{equiv}$. calculate the resistance of a 0.5 M solution of NaCl between plates that are 0.5 cm apart and have areas of 1 cm^2 each.

2. If PbC_2O_4 and CaC_2O_4 are insoluble salts, describe how you could make an electrode reversible to Ca^{2+} ion.

3. A solution contains 1 M NaCl and $CaCl_2$. (a) determine the transport numbers of each ion in terms of the mobilities and charges of each ion. (b) Assume the concentration differences are small, for example Solution 1 = 1.0 M NaCl and 1.0 M $CaCl_2$ and Solution 2 = 1.1 M NaCl and 1.1 M $CaCl_2$, so that the transport number is roughly constant over the interval. Derive the diffusion potential for this system if

$$u_{Na} = 5 \times 10^{-8} \, \text{m}^2/\text{s} \cdot \text{V}$$

$$u_{Ca} = 6 \times 10^{-8} \, \text{m}^2/\text{s} \cdot \text{V}$$

$$u_{Cl} = 7.5 \times 10^{-8} \, \text{m}^2/\text{s} \cdot \text{V}$$

(c) Set up the same problem using the Henderson equation. How much do the two results differ?

4. Use the Henderson equation to develop an expression for the diffusion potential for two solutions containing different concentrations of $CaCl_2$.

5. When 1 M KCl is used as the solution for a salt bridge ($u_K = u_{Cl} = 8 \times 10^{-6}$ m/s), a potential of -27 mV is observed when the connected solutions are 0.5 M and 1.5 M, respectively. When a solution of LiCl is used as the salt bridge, the potential measured is -18 mV. (a) What is the magnitude and sign of the diffusion potential? (b) What is the mobility of the Li^+ ion? Show your steps.

6. A cell with transference and two reversible electrodes gives a potential

$$\psi = (-RT/F)2t_+ [\ln(c_2/c_1)]$$

while the diffusion potential is

$$\psi_d = -(RT/F)(t_+ - t_-)[\ln(c_2/c_1)]$$

(a) Show that the difference between these potentials is

$$\Delta\Psi = -(RT/F)[\ln(c_2/c_1)]$$

(b) Explain the significance of the result from (a) for this univalent–univalent case.

Chapter 6

Electronics

1. INPUT AND OUTPUT IMPEDANCE

When measuring instruments are used to examine a system, the resultant measurements are often made without regard to the effect that this instrument might have on the system it is measuring. During a measurement, the measuring device must draw some current, and this current drain may have a significant effect on the magnitude of the potential itself.

In order to describe this phenomenon more effectively, the measuring device is assumed to be a black box with a pair of input ports. Although the circuitry within the box may be quite complex and involve a number of current pathways, these pathways can always be compressed into a simple circuit such as the resistance–capacitance combination of Fig. 6.1. Because the only knowledge of the black box is acquired through these input ports, the actual circuit detail is not important for this analysis. Most instruments will list the values of the input resistance and input capacitance. For example, an oscilloscope may have 1 MΩ resistance (10^6 Ω) and 47 pF capacitance (1 pF $= 10^{-12}$ F). To examine the role of these parameters, the input resistance will be analyzed. Extensions to an input impedance are then straightforward.

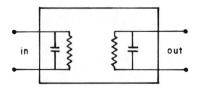

Fig. 6.1 Input and output ports for a measuring instrument.

114

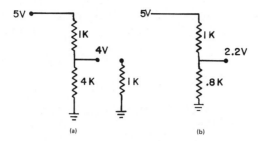

Fig. 6.2 Perturbation of the output by a measuring device. (a) The potential for the unperturbed circuit. (b) The reduced potential produced when 4 KΩ and 1 KΩ are used in parallel during measurement: 0.8 KΩ is the net parallel resistance.

Consider the simple series circuit of Fig. 6.2(a). The potential drop across the lower resistor is to be measured with a measuring device having an input impedance of only 1 KΩ. For the 5 V input, the voltage drop across the 4 KΩ resistor is calculated to be 4 V, because

$$V_2 = [R_2/(R_1 + R_2)]V_{tot} = [4/(1 + 4)](5 \text{ V}) = 4 \text{ V} \qquad (6.1)$$

However, if the values of the resistors were now known and the measuring device with a 1 KΩ input impedance was applied at this point to measure the potential, the 5 V would flow through the modified circuit of Fig. 6.2(b). The 4 KΩ and 1 KΩ resistors combine in parallel to produce an equivalent resistance of 0.8 KΩ. The net voltage drop across these parallel resistors would then be only 2.2 V, so that the measuring device has introduced a sizable error into the potential measurement.

The situation can be ameliorated if the input resistance of the measuring device is increased significantly. For example, if the input resistance were increased to 1 MΩ, the parallel combination of the 4 KΩ and 1 MΩ resistors would produce an equivalent resistance of 3.98 KΩ and the perturbation would be relatively small. The perturbation from the measuring device can be minimized by maximizing the input resistance (or impedance) of the measuring device.

Just as the input to a device can be reduced to a simple equivalent circuit, the output of a device can be reduced similarly. The requirements for the output port are different, however. For many devices, considerable output current is required. For example, a servomotor may have a net resistance of 100–1000 Ω and draw currents on the order of 1 A. The output of the driving unit can be perturbed when large currents are required. To minimize this effect, consider an output resistance of 10 Ω, which drives a device with an input impedance of 500 Ω. If the two resistances are combined in parallel, the resistance is 9.8 Ω. Even though the device draws some current, the bulk of the current still flows through the 10 Ω resistor and the effects of the added device are minimal. Perturbations in the output current are minimized by making the output impedance of the device as low as possible.

2. OPERATIONAL AMPLIFIERS

Operational amplifiers are devices that have the very high input impedances that make them ideal measuring instruments. In operational amplifiers designed with FET (field effect transistor) inputs, the input impedance may be as high as 10^{14} Ω.

The operational amplifiers also have extremely high gain. The gain is defined as the ratio of the output voltage of the device to its input voltage and is a measure of the device's ability to amplify the signal. Operational amplifiers typically have gains of 10^5–10^6. The presence of such gain when no feedback is possible between the input and output is called open loop gain, because the input is connected to the output only through the amplifier. The presence of such high gains presents an interesting limitation of the devices. The output voltages can never exceed the power supply voltages of the device, which are typically 15 V. However, this maximum potential is attained when the input voltage in the open loop configuration is only 150 μV. The device is easily saturated. In addition, the high gain amplifies any small signals, including random noise, that may be present. Because of this, the open loop amplifier is rather impractical.

The advantage of operational amplifiers lies in their use for closed loop configurations, in which the output of the device is connected back to the input in some manner. The voltage amplifier has an inverting input, so that feedback to the input will have a reversed polarity, and also has a positive noninverting input. For signals at the inverting input the feedback potentials can be used to minimize the effects of voltages at the input.

Operational amplifiers always have both positive, or noninverting, and negative, or inverting, inputs. Feedback to a positive input normally tends to reinforce any input signal applied to the amplifier. A typical example of such positive feedback is a microphone–speaker system. If the speakers emit a signal that can be picked up by the microphone, the signal is amplified still further until the system reaches its saturation level. The resultant squeal is eliminated by reducing the gain of the amplifier and, hence, the amount of feedback.

The negative feedback can stabilize the system, because it reduces the effects produced by variations in the input. In fact, negative feedback will act

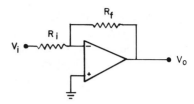

Fig. 6.3 An inverting amplifier. Input current is compensated by current through the feedback loop.

to keep the potential difference between the positive and negative input ports the same. Any deviations between the two ports will produce a feedback that drives the input potential difference back to zero. This fact will provide a simple tool for analyzing the effects of the various operational amplifier circuits utilizing feedback. For example, if the positive input of the operational amplifier is grounded—that is, connected to the chassis or the ground of the power supply—the inputs and feedback to the negative input port must also have ground potential for stability.

This approach is illustrated by a simple example, as shown in Fig. 6.3. A potential is applied to the negative input port through a resistor R_i. The positive input port is grounded and a feedback resistor, R_f, connects the operational amplifier output and the negative input. According to the requirement of equal potentials, the negative input should adjust itself to ground potential. This is possible only if the current flowing in the feedback loop exactly cancels the current flowing through R_i to the input. The currents will have opposite sign, because the amplifier inverts the signal. Equating the two equal and opposite currents,

$$- V_i/R_i = V_{out}/R_f, \qquad (6.2)$$

V_{out} is the voltage of the amplifier output, which can be expressed as

$$V_{out} = -(R_f/R_i)V_i \qquad (6.3)$$

In other words, this voltage, V_{out}, from the amplifier output is sufficient to balance the negative and positive currents. The negative input will assume zero potential because of the cancellation of the two opposing currents; for this reason, it is called a virtual ground because the operational amplifier is actively holding it at ground potential.

Note that no cancellation would take place if, instead of the negative input port, the positive input port received the feedback. Because both input and output potentials have the same sign, they could not cancel and the system would rapidly rise to its saturation level.

When the gain of the operational amplifier is extremely high so that the inputs must remain at the same potential if the system is to be stable, the actual characteristics of the amplifier do not enter. The output characteristics are dictated only by the magnitudes of the two resistors in the circuit. This is a major advantage of such circuits, because each operational amplifier should give the same type of output. The high gain that produces saturation in the open loop configuration becomes a strong advantage in feedback applications by eliminating the effects of different operational amplifiers. Thus, the same type of circuit can be used for any of the commercially available operational amplifiers.

In actual circuits, a small capacitor is often placed in parallel with the feedback resistor to minimize the contribution of very high frequency components that might lead to oscillations. The impedance of this capacitor is

$$X = 1/\omega C \tag{6.4}$$

and is the lowest at the higher frequencies. If this impedance appears in place of R_f at high frequencies where it is the major pathway, the expression for V_{out} is approximately

$$V_{out} = -(1/\omega C R_i)[V_{in}(\omega)] \tag{6.5}$$

and the "amplification" of any high frequency component will become smaller as the frequency increases.

3. OPERATIONAL AMPLIFIER SUMMER

In the previous section, the operational amplifier produced a new output voltage for some input voltage. The gain was $-(R_f/R_i)$, and because the resistors are normally comparable in size, the system is not normally as convenient as a high-gain amplifier.

Other advantages of the operational amplifier negative feedback configuration are apparent in the following problem. Two voltages, $V_1 = 10$ V and $V_2 = 5$ V, are to be added. Any configuration that pits one potential against the other will depend on the net potential difference rather than the actual voltages. For example, suppose the voltages are arranged as shown in Fig. 6.4(a). The circuit reduces to that shown in Fig. 6.4(b), in which a potential of 5 V drops through a total resistance of 2 KΩ. There is no way to get a net potential of 15 V, the sum of the two potentials. The two potentials can be summed, however, by adding all the current from the two of them when

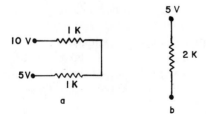

Fig. 6.4 (a) A resistor arrangement that cannot sum 10 V and 5 V, as it reduces to configuration (b) based on the net potential difference and the resistor series sum.

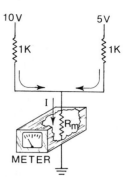

Fig. 6.5 Summing 10 V and 5 V by directing currents generated for each voltage via a resistor through a resistance-free meter to ground potential.

they do not interfere with each other. In Fig. 6.5, such an arrangement is shown. Both potentials are allowed to drop to ground potential. Because they fall to a common potential, each behaves independently. An ideal ammeter, with no internal resistance and no potential drop, is inserted between the summing point and ground and sums all the currents passing to ground. To convert this current to an equivalent voltage, it must be passed through a resistor. Obviously, such a resistor cannot be inserted in the circuit, because it must change the overall current.

An operational amplifier with negative feedback converts the voltages into a sum at its output, and the circuit is shown in Fig. 6.6. No matter how many different inputs are applied to the negative port, the output voltage and feedback resistor will supply sufficient current to cancel these input currents and bring the port to virtual ground if the positive port is at ground potential. In the example of Fig. 6.6, two input currents are balanced by the feedback current, that is,

$$(V_1/R_1) + (V_2/R_2) = -(V_{out}/R_f) \qquad (6.6)$$

Rearranging,

$$V_{out} = -(R_f/R_1)V_1 - (R_f/R_2)V_2 \qquad (6.7)$$

Each input voltage sums independently, and the process could be extended to any number of inputs.

To illustrate use of the summer, let us determine an electrical analog circuit for the expression

$$y = x + 2u + 3v \qquad (6.8)$$

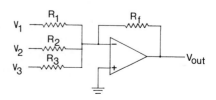

Fig. 6.6 An operational amplifier with negative feedback used as a voltage summer.

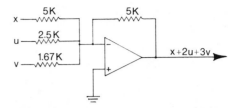

Fig. 6.7 Operational amplifier summer as an analog circuit for the algebraic function $y = x + 2u + 3v$.

In the circuit, x, u, and v will be fed into the negative port using the proper input resistors. If the feedback resistor is chosen as 5 KΩ, the input resistor for x, expressed as some voltage, will be 5 KΩ so that

$$(R_f/R_1) = (5 \text{ K}\Omega/5 \text{ K}\Omega) = 1 \qquad (6.9)$$

The input resistor for the u input must give 2 times the output, so R_2 is chosen as 2.5 KΩ,

$$(R_f/R_2) = (5 \text{ K}\Omega/2.5 \text{ K}\Omega) = 2 \qquad (6.10)$$

Finally,

$$(R_f/R_3) = (5 \text{ K}\Omega/1.67 \text{ K}\Omega) = 3 \qquad (6.11)$$

The entire configuration is shown in Fig. 6.7. Notice that the total output is negative. The potential can be inverted by directing it to a second operational amplifier in which the input and feedback resistors are identical. This configuration will not change the magnitude but will reverse the sign of the potential.

In the summer circuits, an input potential generated current through an input resistor, and this current was then cancelled by the feedback current. Often, an input current must be measured. For such measurements, the input potential and input resistor are replaced by the input current, as shown in Fig. 6.8. The operational amplifier produces an output voltage sufficient to cancel the incoming current at the summing junction, and the output voltage is then a measure of the current reaching the virtual ground. This arrangement is called a current-to-voltage converter and is an extremely useful device

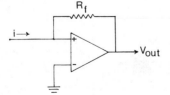

Fig. 6.8 The current-to-voltage $(i–V)$ converter with $i_{in} = -V_{out}/R_f$.

for current measurement. Because the current in the feedback loop must equal the input current, the current of interest is directly related to both the output voltage and the feedback resistance as

$$i = -V_{out}/R_f \qquad (6.12)$$

4. NONINVERTING AMPLIFIER

To record a potential without inversion, the positive input port is used. However, negative feedback is still necessary to maintain a stable configuration. Since both ports are to have the same potential, this feedback must be arranged to permit the negative port to assume the potential of the positive port. The circuit used to do this is shown in Fig. 6.9. The resistance R_1 provides a path to ground while maintaining a finite input potential, because this port cannot be grounded while the positive input potential is finite. If the negative input is assumed equal to V_{in}, the current through R_1 is V_{in}/R_1. Further, V_{out} generates a feedback current through R_f and R_1, so that the current is

$$V_{out}/(R_1 + R_f) \qquad (6.13)$$

The currents must be the same for both inputs, so

$$V_{in}/R_1 = V_{out}/(R_1 + R_f) \qquad (6.14)$$

or

$$V_{out} = [(R_1 + R_f)/R_1]V_{in} = (1 + R_f/R_1)V_{in} \qquad (6.15)$$

Although the noninverting circuit has some similarities with the inverting circuit, some differences are worth noting. In the inverting circuit, both

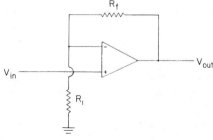

Fig. 6.9 The noninverting operational amplifier configuration.

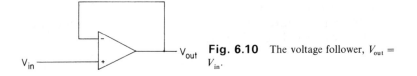

Fig. 6.10 The voltage follower, $V_{out} = V_{in}$.

inputs were held at the same potential. For the noninverting configuration both inputs are held at the same potential, but this is the applied input potential, which is not normally ground.

The potential is applied directly to the positive input port. For this reason, it encounters a very high input resistance, which, as noted previously, may approach $10^{14}\,\Omega$. For the inverting amplifier, the input resistance was effectively the resistance of the input resistor. Because of its extremely high input resistance, the noninverting configuration is often used to buffer some potential from the rest of the circuit. The circuit draws little input current and produces a new voltage proportional to the original. This new voltage can then be used to drive subsequent circuitry.

The expression for V_{out} cannot be reduced to unity when finite values of R_1 and R_f are used. The factor $(1 + R_f/R_1)$ will reduce to unity if R_f is zero or R_1 is infinite. In practice, both these conditions are satisfied by creating the configuration of Fig. 6.10, which is called a voltage follower because the output of the circuit follows the input voltage while buffering it from subsequent circuitry. The negative input and output are shorted to produce $R_f = 0$, while the resistor R_1 is eliminated completely to create an infinite resistance pathway to ground. While the input impedance of this circuit is extremely high, the output impedance is quite low because of the zero feedback resistance. The circuit can be used directly to drive systems requiring high currents with minimum distortion of the input voltage.

Both the inverting and noninverting inputs can be used to make a difference amplifier for measuring the difference between two potentials. A difference amplifier is essentially a combination of the inverting and noninverting configurations. To create a difference, one of the two signals must be inverted. For this, the negative input and a standard inverting feedback configuration are used, as shown in Fig. 6.11. The second potential might now be fed directly into the positive input, but this configuration would simply bring the negative summing junction to the same potential and would not produce a potential difference. The configuration that produces differences is shown in Fig. 6.11. The positive input is fed by a voltage divider configuration, so that the input to the summing junction is

$$v_2' = [R_2/(R_1 + R_2)]v_2 \qquad (6.16)$$

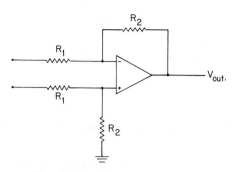

Fig. 6.11   The difference amplifier.

For convenience, equal pairs of resistors are selected for the positive and negative summing junctions. The voltage v_2' is the positive input. The potential v_1 will give an output

$$V_{out} = -(R_2/R_1)v_1 \tag{6.17}$$

The input voltage v_2 is converted into v_2', which in turn is converted to

$$V_{out} = [(R_2 + R_1)/R_1]v_2'$$
$$= [(R_2 + R_1)/R_1][R_2/(R_1 + R_2)]v_2$$
$$= (R_2/R_1)v_2 \tag{6.18}$$

When the two potentials of opposite sign add, they produce the final result

$$V_{out} = (R_2/R_1)(v_2 - v_1) \tag{6.19}$$

and the output is proportional to the absolute difference of the two voltages.

5. OPERATIONAL AMPLIFIER INTEGRATORS

In the examples of previous sections, capacitors were introduced only to the operational amplifier feedback; for this reason, they were generally quite small. If a larger capacitor is used in the feedback loop in place of the feedback resistor, the resultant operational amplifier output is quite different. Consider the circuit of Fig. 6.12. Although the capacitor blocks a direct flow of current from the output to the summing junction, this does not mean that currents will not flow. Instead, they must flow to charge the capacitor. A current, V_{in}/R_1, is flowing into the summing junction. If the applied potential is positive, this current can induce a negative potential at the output.

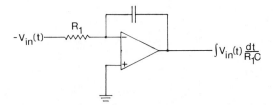

Fig. 6.12   The operational amplifier integrator, $V_{out} = -\int_0^t V_{in}(t)/\tau \, dt$.

Instead of flowing back to the input, this potential drives negative charge to the capacitor where it is stored. Because both sides of the capacitor must charge, the "positive" charge reaching the summing junction simply continues on to the capacitor plate on the input side, where it is collected until the capacitor saturates. If the positive input is held to ground potential, the negative input must also seek this potential and the input currents and capacitor charging currents must be equal, such that

$$i = V_{in}/R_1 = -C \, dV/dt \qquad (6.20)$$

or

$$dV/dt = -V/R_1 C = -V/\tau \qquad (6.21)$$

where $\tau = RC$ is the time constant for the resistor–capacitor series combination. Note that the input voltage may be time dependent. The differential equation can be integrated to give

$$V_{out} = -\int_0^t [V_{in}(t)/\tau] \, dt \qquad (6.22)$$

The circuit integrates the input voltage $V_{in}(t)$ until the capacitor is completely charged. We choose R_1 and C such that the time constant τ is long compared to the time allotted for the integration, to prevent complete capacitor charging.

A differentiator circuit can be constructed by simply reversing the positions of the capacitor and resistor, as shown in Fig. 6.13. In this case, the input voltage charges one plate of the capacitor in the input. Although the process

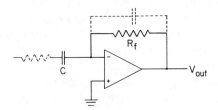

Fig. 6.13   The operational amplifier differentiator, $V_{out} = -\tau \, dV_{in}(t)/dt$.

might be expected to proceed instantaneously, because there is no resistance in the input circuit, both sides of the capacitor require equal and opposite amounts of charge. The feedback resistor controls the flow to the one side of the capacitor and thus controls the overall flow of charge. The operational amplifier "holds" the negative input at virtual ground, and the two current flows to this point are equal and opposite. Therefore,

$$C dV_{in}(t)/dt = -V_{out}(t)/R \qquad (6.23)$$

and

$$V_{out} = -RC \, d[V_{in}(t)]/dt \qquad (6.24)$$

The amplifier output is directly proportional to the derivative of the input potential function.

The differentiator circuit is not used as widely as the integrator circuit because it is inherently noisy. While the integrator collects and smooths fluctuations in the input potential, the differentiator tends to enhance these fluctuations, particularly at higher frequencies. The integrator can be used as a low-pass filter that preferentially passes the lower frequencies. By comparison with the summer gain equation,

$$A = (R_f/R_1) \qquad (6.25)$$

the gain of the integrator will be

$$A = (1/\omega C)/R_1 = 1/(\omega R_1 C) \qquad (6.26)$$

and the gain of the circuit decreases at higher frequencies. The differentiator has a gain of

$$A = R_1/(1/\omega C) = \omega RC \qquad (6.27)$$

and its gain is maximal at the higher frequencies. It will tend to magnify sharp (high-frequency) variations in the signal.

The differentiator circuit is extremely convenient for the generation of trigger pulses. For a square voltage pulse, the derivative is zero for regions of constant potential and becomes large only during the transition phases of the square pulse. A differentiator circuit will produce spikes both at the beginning and end of the square wave, as shown in Fig. 6.14. These spikes may then be used to trigger circuitry that must start at the same time as the square pulse.

The integrator circuit can be used to generate a ramp potential, which could then be used to drive the x axis of an x–y recorder at some constant rate. Such a ramp is produced when a constant input potential V is applied to the integrator. The output voltage is

$$V_{out} = -(V/RC) \int_0^t dt = Vt/RC \qquad (6.28)$$

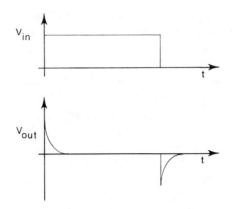

Fig. 6.14   Differentiation of a square voltage pulse.

The ramp slope is controlled by the values of R and C and gives a result such as that shown in Fig. 6.15. A shorting switch is often added to the feedback loop to discharge the capacitor after a given integration.

In this discussion, the integrator and differentiator circuits have been assumed ideal. In actual circuits, additional components must be added to stabilize the system. For example, a small resistor is often added to the feedback loop of the integrator to bleed some of the current. Obviously, this resistor slows the response of the system and the resultant integration is not as accurate, but the system remains more stable. When the integration is complete, this resistor provides a pathway for discharge of the capacitor. The feedback loop often includes a shorting branch with a switch as well. On completion of an integration, the capacitor is shorted to discharge and reset the integrator for the next experiment.

In the differentiator, a small resistor is often included in series with the input capacitor, because the resistor provides extra resistance for the high-frequency components as well as the low and tends to reduce the noise contributions of high-frequency components. In actual integrator and differentiator circuits, there may actually be resistors and capacitors in both the

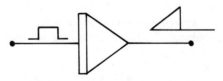

Fig. 6.15   A potential ramp produced with a constant potential input to an integrator.

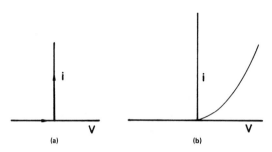

Fig. 6.16 (a) An ideal current versus voltage curve for a diode. (b) The current versus voltage exponential curve for a semiconductor diode.

input and feedback loops, but these components will be small relative to the major components.

In the analog circuits discussed thus far, the circuit elements have all been linear elements—that is, resistor and capacitors. However, the elements do not have to be linear, and some interesting results appear when a diode is used in the circuits.

The ideal diode has infinitely high resistance for negative voltages and zero resistance for positive voltages, as shown in Fig. 6.16(a). In reality, the current–voltage curve for a diode is unable to make such a sharp transition and is exponential, as shown in Fig. 6.16(b). We can take advantage of this exponential dependence in operational amplifier circuits. Consider a case where the diode is inserted in the feedback loop of an operational amplifier, as shown in Fig. 6.17(a). The diode is arranged to permit positive current to flow from left to right and the amount of current will increase exponentially with increasing potential. At the lower voltages, the resistance of the diode is high and the gain of the device is high. As the voltage increases, the diode resistance drops and the gain also decreases. The resultant output is proportional to the logarithm—that is, the inverse of the exponential—of the input voltage, and the diode acts to compress the voltage scale. Large voltages produce proportionately lower voltage outputs because of the reduced gain of the amplifier. Such circuits are extremely useful where experimental data scan many orders of magnitude, such as the current outputs from a photomultiplier.

As might be expected, a diode in the amplifier input circuit will enhance the voltage gain at higher potentials, because more current can enter the summing junction, and will decrease the gain at the lower potentials. The circuit is shown in Fig. 6.17(b), and the potential output will have the form

$$V_{out} = -(\exp a) V_{in} \tag{6.29}$$

where a is a constant, characteristic of the diode and resistor used.

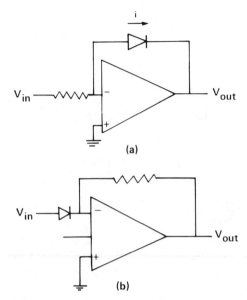

Fig. 6.17   (a) A logarithmic amplifier. (b) An exponential amplifier.

6. VOLTAGE CLAMP

Many of the operational amplifier circuits with negative feedback could
be analyzed by simply noting that the feedback was used to bring both input terminals of the amplifier to a common potential. If the noninverting
input were grounded, the inverting input would be held at a virtual ground.
This property serves as the basis of a control system. Consider a nerve axon
in its resting (equilibrium) state with a potential of -60mV. There is no net
current through the membrane. What operational amplifier configuration
might be used to reflect this fact?

Consider the simple circuit of Fig. 6.18. A positive 60 mV potential
is connected to the inverting input through a 5 KΩ resistor. A second 5 KΩ

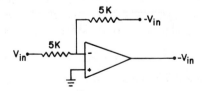

Fig. 6.18   A basic voltage clamp system.
Balanced circuit—feedback voltage introduced independently.

resistor also enters the summing junction. If this resistor were connected to the output, then a simple inverter circuit would result. The operational amplifier would generate a potential of -60 mV to maintain the virtual ground. However, this virtual ground could also be maintained if a potential of -60 mV from the axon were applied to the 5 KΩ resistor directly, as shown in Fig. 6.18. Because the two input potentials are identical, no current would flow in the output circuit of the operational amplifier.

If the independent feedback voltage were selected as the potential difference across the membrane, then this voltage would have to be identical to the control voltage for the operational amplifier to be balanced. This is the operating principle of the voltage clamp, which is shown in Fig. 6.19. Control potentials are applied through input resistors. The control pulse, which must be negative to produce a positive potential across the membrane, is applied to one input while a constant potential equal to the negative of the cell resting potential is applied through a second resistor. This second input ensures a transmembrane potential equal to the resting potential when the system is in clamp without an applied control pulse. A control pulse will then raise the transmembrane potential by some potential—such as 80 mV— relative to the resting potential: that is, $-60 + 80 = +20$ mV. If the system is balanced, the voltage across the nerve membrane must also be $+20$ mV.

With the "disconnected" feedback arrangement of the voltage clamp, we must reexamine the output of the operational amplifier. The actual potential difference that is measured for the feedback loop must be generated by the iR drop across the membrane, solution, and so on. If the output of the operational amplifier is directed into the membrane via a current-carrying electrode, current will flow to produce an iR drop through the membrane equal to the control potential. This current is collected by external electrodes and converted to a calibrated potential (mA/cm^2) with a current-to-voltage converter. In other words, we measure the current necessary to maintain the axonal potential difference equal to the control potential as a function of time.

Because the operational amplifier of the voltage clamp circuit functions as a summer, a variety of potential inputs can be applied to the axon simultaneously. For example, two clamp pulses might be applied in sequence by using two separate input lines. The first potential pulse is applied through the first line while a delayed pulse is applied through the second. The potential drop established by the control pulses is the iR drop across the membrane plus all additional resistive components. Because we are interested in only the potential drop across the membrane, it is convenient to remove the remaining potential component. This series resistance compensation can be done by feeding back a potential that is proportional to the total current

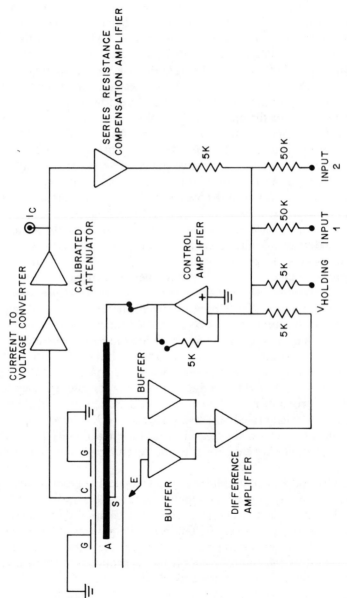

Fig. 6.19 A complete voltage clamp system.

through the membrane. The arrangement is shown in Fig. 6.19. The voltage from the current-to-voltage converter is directly proportional to the current through the region between the internal and external current electrodes. If this voltage is applied to an inverting amplifier, the operational amplifier will produce a potential of opposite sign that can be applied to a summing input of the control amplifier. The magnitude of the feedback element can be adjusted to provide different series resistance compensations. The compensating potential will then increase the transmembrane potential so that the potential across the membrane alone is equal to the control potential. For example, a control potential of 60 mV may dissipate 3 mV in the series resistance so that the membrane "sees" a potential of only 57 mV. The positive feedback of the series resistance branch will increase the total transmembrane potential to 63 mV so that 60 mV is now dissipated in the membrane.

The inputs to a voltage clamp need not be square pulses. The axon can be clamped with a variety of potential functions, such as ramps and hyperbolas. If the axon is stimulated to produce an action potential, this action potential can be stored in a computer memory. This stored action potential is then reconverted to a potential signal and used as the input for an action potential clamp experiment. The current observed reproduces the stimulating current, which generates the original action potential, as shown in Fig. 6.20.

The action potential clamp experiments were performed using an electronic analog, the Lettvin axon, which mimics the behavior of normal biological axons. A Lettvin axon analog circuit is illustrated in Fig. 6.21.

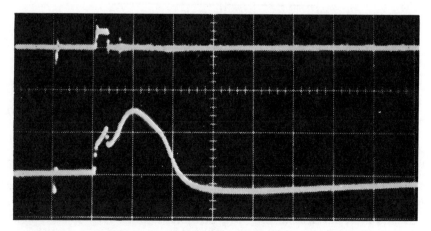

Fig. 6.20 Action potential clamp for the Lettvin analog axon. Upper trace: current record produced with the stored action potential as the control potential. Lower trace: instantaneous potential of the Lettvin analog axon during clamp. Scale, 100 mV/div, 1 ms/div.

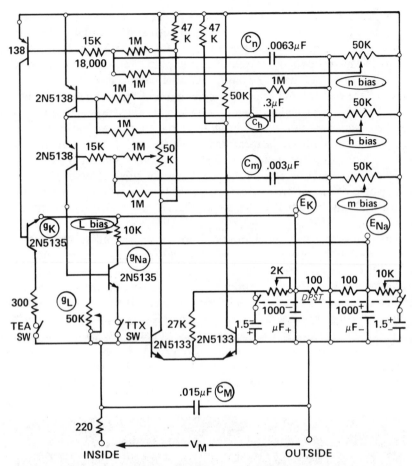

Fig. 6.21 The Lettvin analog axon—all transistors Fairchild. [Reproduced with kind permission of Fishman, pers. com. 1976.]

7. LOGIC CHIPS

The circuits constructed using operational amplifiers are called linear circuits, as they can assume any potential output. An alternative family of circuits are built around only two possible outputs; these are the digital circuits. Such circuits will have either a high output (often designated as a "1") and a low output ("0"). All applied input potentials that fall below some potential level will produce a common low output, while potentials

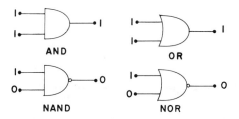

Fig. 6.22   The AND, NAND, OR, and NOR gates.

above this level will produce the high output. Although a variety of high and low potentials are used, we will focus on TTL (transistor–transistor logic) systems where potentials below approximately 1 V are classified as low and those above this level are high. The low output is 0 V while the high output is 5 V.

The digital equivalents of the summer circuits are the AND, NAND, OR, and NOR gates, which are illustrated by the symbols of Fig. 6.22. The AND gate will give a high output only when all inputs are high. If any one of these inputs goes low (0 V), the output will also drop to zero. The NAND gate is similar. If all the inputs are high, the output is low (Not-AND). If one or more inputs go low, the NAND gate output goes high.

The OR and NOR gates will go high and low, respectively, when any or all of the inputs go high. An exclusive OR represents a special case, in which the output goes high if either of two inputs goes high but will remain low if both inputs are high.

The logic circuits can couple to linear circuits when the absolute magnitude of an event is not important, but we must have information that events are taking place. For example, we might wish to trigger detection circuitry when an axon is voltage clamped and light from a flashlamp impinges on the axon. Figure 6.23 shows an AND used for this purpose. When both the amplified voltage clamp control voltage and the flashlamp trigger outputs are greater than 1 V, a 5 V trigger pulse will trigger the detection circuitry.

Some logic chips are listed as open collector circuits. In such cases, an output is observed only if a 5 V potential and resistor are included in the output circuit. In the high state, the resistance of the logic chip is very high so that the potential at the input to the chip is approximately 5 V—that is,

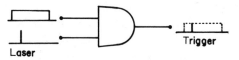

Laser

Trigger

Fig. 6.23   An AND gate used as a trigger.

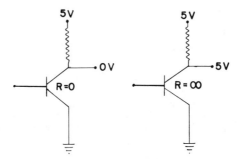

Fig. 6.24 An open collector logic system.

little potential is dissipated in the external resistor. In the low configuration, the chip becomes an open circuit, most of the voltage drops across the external resistor, and the voltage at the chip is zero, as shown in Fig. 6.24.

A comparator logic chip is used when we wish to be more discriminating in our choice of potential that will trigger a high output. The comparator has both positive and negative inputs. When the potential at the positive input is larger than that of the negative input, the circuit will give a high output. It gives a low output when the positive input is less than the negative input.

The comparator can be used as a sensing device. For example, we might wish to turn off a voltage clamp if the transmembrane potential exceeded some preset value, such as 200 mV. A 200 mV potential is then applied to the negative input, while the positive input is connected to the output of the difference amplifier which measures the transmembrane potential. If the transmembrane potential exceeds 200 mV, the resultant 5 V output can be used to trigger a switch that turns off the voltage clamp.

The circuits shown here follow the input pulses. If the two pulses to the AND gate are each high for x seconds, the output will remain high for these x seconds as well. In many cases, however, it is necessary to stretch a pulse. This can be done with a monostable multivibrator chip. An RC circuit applied externally to the chip dictates how long the monostable chip will remain on, and this is completely independent of the duration of the trigger pulse.

The monostable is used when we wish to delay one signal relative to another signal. For example, we may wish to trigger a flashlamp at some time after application of a control potential in a voltage clamp experiment. The RC components of the monostable chip would be set for the proper time, and the monostable chip would be triggered by the rising phase of the control potential. When the monostable chip turned off, the transition from high to low could be used to trigger the flashlamp.

8. OTHER INTEGRATED CIRCUIT MODULES

With the continued development of microcircuits, more distinct circuits can be etched on a single chip. An AND gate chip, for example, can contain three or four separate AND gates for a typical 14-pin chip. With this flexibility, it is also possible to include more complicated circuits on a single chip.

The 555 timer chip is used for a variety of timer applications. The chip will produce an output pulse when triggered like the monostable, but it can also be programmed to produce a continuous series of pulses. Using a proper choice of resistors and capacitors, high and low periods during a given cycle (the duty cycle) can be controlled. Thus, the timer can produce a continuous train of pulses with a preset duration and a preset time between pulses. Figure 6.25 shows a 555 timer used to drive a stepper motor controller. The speed of rotation of the stepper motor is dictated by the frequency of pulses from the timer. A switch selects a resistor for fast "coarse" scanning, while a resistor–potentiometer combination permits a slow scan that can be set very accurately. The circuit is used to scan wavelengths of a tuneable dye laser.

Function generators are used to generate sine or square waves at different frequencies. The circuitry for an entire function generator can be developed on a single chip such as the EXAR 2206. The frequency of the unit is regulated by external resistors and capacitors, and the unit will produce sine, triangle, and square wave output.

Voltage regulators are used to produce a specific output voltage when an input voltage greater than this output voltage is applied to the input. Such

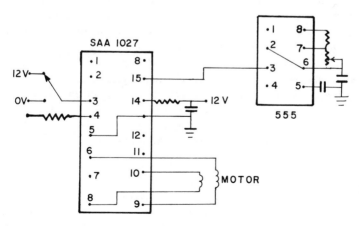

Fig. 6.25 A stepper motor controlled by a 555 timer chip and SAA1027 controller chip.

regulators are useful when a circuit contains a mixture of linear and digital components. Operational amplifiers require ± 15 V inputs, while the logic circuits require 5 V. A single ± 15 V supply can be used for both types of circuit if the $+ 15$ V output of the power supply is applied to the input of a 5 V regulator. Regulators are available for most of the standard voltages (± 5, ± 6, ± 12, ± 15). In addition, precision voltage regulators are available. Such regulators are used when a voltage must be known and maintained with high accuracy.

9. SIGNAL AVERAGING

In many experiments, the signal of interest is extremely small and becomes comparable to noise that is generated in different portions of the instrumentation. In order to extract this signal from the noise, we can take advantage of the fact that the signal itself should be reproducible from experiment to experiment while the noise should be random. If the total outputs for each experiment are added, the noise will cancel while the signal will grow. By adding enough experiments, the signal should rise above the noise level for easy detection. This is the basic principle of signal averaging.

To average a signal, it is necessary to store each experiment and then add them. The most flexible way to do this involves converting the analog voltage signal into a digital form using an analog to digital converter. The digital form of the analog voltage is a sequence of highs and lows (ones and zeros) that can be stored in a computer memory. To illustrate the conversion process, we describe a successive approximation converter. Consider a four-bit converter that can convert an analog signal into a string of four binary digits: for example, 1111 might correspond to a maximum voltage of 10 V, 0000 would correspond to 0 V, and 1000 would correspond to 5 V. Obviously, any voltage over 5 V would give a digital number over 1000. The successive approximation converter samples the signal in a series of steps. for example, let us digitize 3 V using the converter. The converter will first determine if the voltage is greater than or less than 5. In this case $3 < 5$, so a zero is placed in the first (left-hand) location of the four-bit digital number,

$$0\_\_\_ \tag{6.30}$$

The converter has now determined that the voltage is between 0 and 5 V. In the next step, it determines its location relative to 2.5 V. In this case, $3 > 2.5$, so a 1 is inserted as the second digit of the binary number. The number is between 2.5 and 5. The third bit is a 0 because 3 V is less than 3.75 V, the midpoint of 2.5 and 5 V:

$$0 \ 1 \ 0 \_ \tag{6.31}$$

The final bit is determined around a center of 3.125. Because 3 is less than 3.125, the digitized number for 3 V is

$$0 \ 1 \ 0 \ 0 \tag{6.32}$$

where 0100 could stand for any voltage between 2.5 and 3.125. To increase the accuracy, it is necessary to add additional digital bits to the number. The four-bit analog-to-digital (A/D) converter will resolve fractions of the order

$$1/2^4 = 1/16 \tag{6.33}$$

Thus the voltage is broken into 16 separate regions. For 10 V, these regions are 0–0.625 V, 0.625–1.250 V, 1.250–1.875 V, and so on. An eight-bit converter will divide the voltage into

$$2^8 = 256 \tag{6.34}$$

equal parts.

Conversion of data into digital form is only one part of signal averaging. The signal will change as a function of time. To store the signal, it must be digitized at regular times during the signal period. The digital numbers for each time interval must then be stored in successive memory locations. The speed of A/D conversion is limited by the time required for the system to convert each signal into digital form. The more common signal averagers have conversion times of 10–20 μs.

To signal average, the averager is started each time a signal is to be observed. The averager digitizes a series of voltages for the preselected time intervals and stores these in a memory. The experiment is then repeated and the new digital numbers are added to those already in the appropriate memory locations. For a total of N such averages, the signal will rise relative to the noise by the factor $\sqrt{N}$; that is, 100 signals must be averaged to improve the signal by a factor of 10 relative to the noise.

To observe the averaged signal, the digital information in the memory must be converted back to analog form. A digital-to-analog (D/A) converter performs this task. The resultant signal can then be displayed on an oscilloscope.

An easily constructed signal averager (Basano and Ottonello, 1981) is illustrated in Fig. 6.26. The circuit illustrates the modular nature of electronics. The signal enters a sample-and-hold module. When the module is triggered by a clock, it "holds" the voltage sensed at that instant. This "constant"

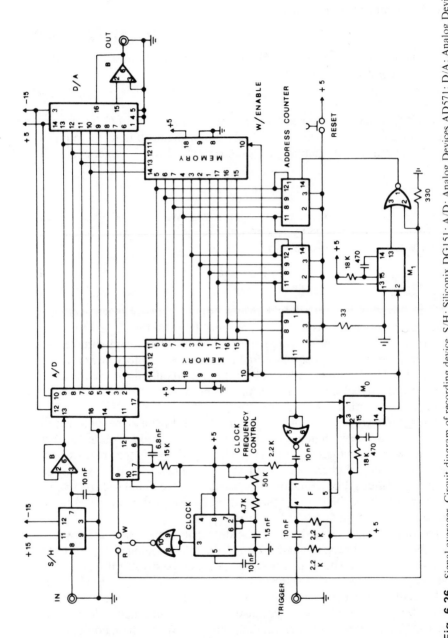

Fig. 6.26 Signal averager. Circuit diagram of recording device. S/H: Siliconix DG151; A/D: Analog Devices AD571; D/A: Analog Devices AD561; B: PMI OP-01; A/D: Analog Devices AD571; D/A: Analog Devices AD561; B: PMI OP-01; Memory: 2x Intel 2114; Clock: Ne555; F: SN7474; M_0, M_1: SN74123; Address Counter: 3xSN7493. [Reproduced with kind permission of Ottonello, 1981.]

voltage is then converted to digital form by the A/D converter. If the voltage could change during the conversion process, the resultant digital number would be inaccurate. The converted signal appears as eight separate outputs corresponding to the eight digital bits produced by the analog voltage. These eight bits are stored in eight memory locations within two random access memories (RAMs). The same clock that controls the sample-and-hold is used to move the memory through successive groups of eight memory locations. For the circuit shown, there are 1024 memory locations of four bits each for each memory chip.

To read out the signal, the clock is used to scan through the memory locations. For each location, the proper combination of binary digits reappear as output and they are directed to the eight parallel inputs of a D/A converter to reproduce the averaged analog signal.

10. LOCK-IN AMPLIFICATION

The signal averager can be used to detect any signal that repeats itself in some regular fashion. However, there are other techniques for detecting signals buried in noise. For example, if the signal of interest changed at some constant frequency, it should be possible to build a detector sensitive only to that frequency. Because noise appears at all frequencies, much of this noise will not be detected by the detector and the signal should become larger relative to the noise.

A lock-in amplifier is a device designed to detect only signals that are at the proper frequency and provide a direct current output proportional to the amplitude of this frequency component. Its operation is best described by the simple circuit of Fig. 6.27, which is used to detect small changes in

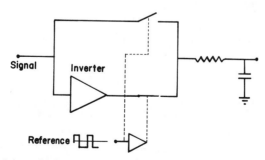

Fig. 6.27   Schematic diagram of the student lock-in amplifier. [Reproduced with kind permission of Temple 1975.]

dye fluorescence caused by changes in the transmembrane potential (Temple, 1975). The sinusoidal function generator that produces the time-varying transmembrane potential also drives the reference input of the lock-in amplifier. The fluorescence signal from the experiment enters the lock-in, is split into two branches, and is inverted in one of the branches. The two signals are recombined after passing through switches in each branch that open and close alternately. The switching time for the switches is determined by the reference signal. For example, a 100 Hz sinusoid would have a total time of 10 ms/cycle. Switch 1 would be open the first 5 ms, while Switch 2 would be open the remainder of the cycle. Consider the situation of Fig. 6.28(a), in which no signal (only noise) enters the lock-in. Random noise will pass through each branch alternately. The resultant output will be averaged with an RC filter in the output line, and this average should approach zero.

When a signal exactly in phase with the reference signal enters the lock-in, the lock-in acts as a rectifier, as shown in Fig. 6.28(b). Switch 1 in the non-inverting branch opens just as the positive portion of the input sine function arrives. As the signal sine function approaches zero, Switch 1 opens and Switch 2 closes to conduct the signal through the inverter so that it also reaches the output as a positive signal. The RC filter in the output integrates these two signals to give a net positive signal, as shown.

The system is optimal when both reference and input signals are exactly in phase or 180° out of phase. The 180° case will give a maximal negative output. If the input signal is only 90° out of phase, no DC signal can be observed, as shown in Fig. 6.28(c). Switch 1 turns on when the input signal is at its maximum. During the time Switch 1 is on, the signal passes through zero and reaches its maximum negative value. During the switch period, equal areas of positive and negative signal pass into the RC output filter and give a net signal of zero. Thus, the phase between the input and reference signals must be adjusted to 0° or 180° to get the maximal output. Many lock-in systems come with a phase control that permits an optimal match between the reference and the input.

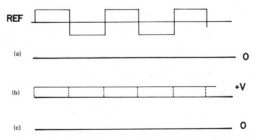

Fig. 6.28 Signal processing in a lock-in amplifier. (a) Reference signal but no input signal. (b) Signal in phase with reference. (c) Signal 90° out of phase with reference.

Chapter 7 _____

Membrane Capacitance

1. RESISTANCE–CAPACITANCE NETWORKS

Membranes display a variety of electrical properties that suggest that they can be modeled as some combination of electrical analog components. For example, an applied potential can produce a current flow through the membrane and this indicates that the membrane has a finite, albeit large, resistance. In addition, the charged groups at the membrane–water interface provide sites for the redistribution of charge at the membrane surface when the transmembrane potential is changed. This charge redistribution is similar to that observed for capacitors in which a change in the applied potential produces a change in the net charge on the capacitor plates. In addition, the bathing solutions also have a resistance. These properties of the bathing solutions–membrane combination suggest that it can be described electrically by the equivalent circuit illustrated in 7.1, where C_m and R_m are the membrane capacitance and resistance, respectively, and R_s is the total resistance of the bathing solutions and electrodes.

Although our discussion will focus on the circuit of Fig. 7.1, more involved equivalent circuits can be proposed to describe the membrane system. Figure 7.2 shows a circuit that includes both the capacitance and resistance of the solutions and stray capacitance possible in the circuit. Other circuits may distinguish different types of membrane capacitance, such as changes in capacitance produced by adsorption or desorption of ions at the surface.

As the circuit of Fig. 7.1 describes the electrical properties of the membrane,

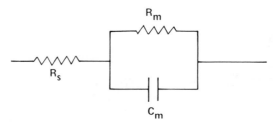

Fig. 7.1   Equivalent resistance–capacitance (RC) circuit for a membrane.

consider the response of this circuit to an applied potential. To simplify the circuit for detailed analysis, assume that the membrane resistance is extremely large, so the capacitative component is the dominant membrane component. The simplified circuit is illustrated in Fig. 7.3. Consider the response of this circuit when the transmembrane potential is increased to a potential V from $V = 0$.

The voltage V must be dissipated across the resistor R_s and capacitor C_m connected in series. To determine the voltage dissipated by each, the total voltage is equated to the sum of the potential drops across each component. The voltage drop in the resistor is

$$V_R = iR_s \qquad (7.1)$$

where i is the current through the resistor. The capacitor is a charge storage device with capacity determined via the relationship

$$C_m = q/V \qquad (7.2)$$

where q is the total charge stored on one plate of the capacitor. The opposite plate must have an equal and opposite charge. The potential drop across the capacitor is then

$$V_C = q/C \qquad (7.3)$$

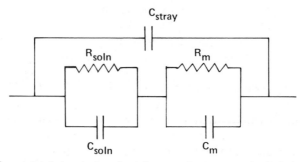

Fig. 7.2   A detailed equivalent circuit for a membrane system that includes electrode and stray capacitances.

Fig. 7.3   A series RC equivalent circuit to test membrane response to an applied potential.

and the total potential drop for the circuit of Fig. 7.3 is

$$V = V_R + V_C = iR_s + q/V \qquad (7.4)$$

In order to solve this equation, we must note that current is charge per unit time, that is,

$$i = dq/dt \qquad (7.5)$$

and the equation becomes a differential equation,

$$V = R_s (dq/dt) + q/C \qquad (7.6)$$

with

$$V = 0 \qquad \text{at} \quad t = 0 \qquad (7.7)$$

and

$$V = V \qquad \text{for} \quad t > 0 \qquad (7.8)$$

The equation can be rearranged to the form

$$dq/dt + q/R_s C_m = V/R_m \qquad (7.9)$$

The product $R_s C_m$ has the units of time and is called the time constant τ for the resistor–capacitor combination,

$$\tau = R_s C_m \qquad (7.10)$$

In order to solve this differential equation, the integrating factor approach is used. If each term of the equation is multiplied by an exponential function of the form

$$\exp(+t/\tau) \qquad (7.11)$$

then the two terms on the left side may be combined as a single differential expression,

$$(dq/dt)[\exp(t/\tau)] + (q/\tau)[\exp(t/\tau)] = d/dt\{q[\exp(t/\tau)]\} \qquad (7.12)$$

Both sides of the equation are now easily integrated with respect to dt:

$$d/dt\{q[\exp(t/\tau)]\} \, dt = (V/R)_s)[\exp(t/\tau)] \, dt \qquad (7.13)$$

Integrating each side from $t = 0$ to some time t produces

$$\{[q(t)][\exp(t/\tau)] - [q(0)][\exp(0)]\} = (V/R_s)[\exp(t/\tau) - \exp(0)] \qquad (7.14)$$

If no charge was present on the capacitor at $t = 0$, $q(0) = 0$. The exponential on the left can be eliminated by multiplying both sides by $\exp(-t/\tau)$ to give

$$q(t) = (VR_sC_m/R_s)[1 - \exp(-t/\tau)]$$
$$= VC_m[1 - \exp(-t/\tau)] \tag{7.15}$$

This time-dependent charge may now be used to determine the potential V_C on the capacitor as a function of time. Because

$$V_C = q(t)/C_m \tag{7.16}$$

the time-dependent voltage is

$$V_C = V[1 - \exp(-t/\tau)] \tag{7.17}$$

At time zero, the potential on the capacitor is zero because charge has not yet reached the capacitor. As time goes on, the capacitor will continue to accumulate charge until the potential reaches V. At this point, the capacitor voltage will balance the applied potential and no further current will flow. The time to reach this level is dictated by the time constant τ, which depends on both capacitor and resistor size. This is physically consistent. A capacitor of larger capacity would require a longer time to acquire its full complement of charge. A large resistor would lower the rate at which charge could flow to the capacitor plate, thus extending the charging time.

The potential across the resistor as a function of time is

$$V_R = iR_s = R_s\,(dq/dt) \tag{7.18}$$
$$V_R = R_s(VC_m)[+\exp(-t/\tau)](1/\tau)$$
$$= V[\exp(-t/\tau)] \tag{7.19}$$

Initially, the full potential drop would fall across the resistor. As the capacitor charged, the portion of voltage across the resistor would drop, reflecting the decreasing current flow in the resistor.

Our series RC circuit was solved rather easily under the condition of a constant applied voltage. Consider now the case in which $R_S = 0$ and the membrane is approximated by the equivalent parallel circuit of Fig. 7.4. If a constant voltage was applied to this system, the potential drop across the two components would just be V. However, different currents i would flow in each branch of the circuit. In fact, the capacitor C would charge immediately because there is no series resistor to limit the flow charge.

In order to examine the time response of this parallel circuit, we apply a constant current i, which must divide between the two branches of the circuit, such that

$$i = i_{C_m} + i_{R_m} \tag{7.20}$$

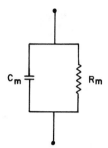

Fig. 7.4 A parallel RC equivalent circuit for membrane response to an applied current.

The current flow will generate a potential difference V. The resistor current is

$$i_{R_m} = V/R_m \tag{7.21}$$

while the current in the capacitative branch is

$$q = CV \tag{7.22}$$

$$i_C = dq/dt = C\, dV/dt \tag{7.23}$$

Equation (7.20) is now a differential equation for V,

$$i = C_m(dV/dt) + V/R_m \tag{7.24}$$

The time constant is generated by dividing all terms by C_m,

$$i/C_m = dV/dt + V/\tau \tag{7.25}$$

For this case, the integrating factor is

$$\exp(t/\tau) \tag{7.26}$$

and the equation becomes

$$\frac{d\{[V(t)][\exp(t/\tau)]\}}{dt}\, dt = (i/C_m)[\exp(t/\tau)] \tag{7.27}$$

Integrating for a constant current that starts at $t = 0$, we find

$$V(t) = R_m i[1 - \exp(-t/\tau)] \tag{7.28}$$

For this parallel circuit, the potential is zero at $t = 0$ because the current flows directly to the capacitor and is unable to develop a potential difference; that is, the capacitor looks like a short circuit to the initial current. As the capacitor is charged at longer times, the equation reduces to

$$V = R_m i \tag{7.29}$$

The capacitor is fully charged, and current flows only through the resistive branch.

The time-dependent currents through each branch are determined from Eqs. (7.21) and (7.23),

$$i_{R_m} = V/R_m = i[1 - \exp(-t/\tau)] \qquad (7.30)$$

and

$$i_{C_m} = C\,dV/dt = RC(i/\tau)[\exp(-t/\tau)] \qquad (7.31)$$
$$= i[\exp(-t/\tau)]$$

The current to the capacitor decays with time as the capacitor charges, while the current in the resistor rises to a maximum as the capacitative current decreases.

2. SINUSOIDAL INPUTS FOR EQUIVALENT CIRCUITS

For the basic series and parallel equivalent circuits for the membrane, the actual time-dependent response of the circuit could be calculated directly. This becomes progressively more difficult as the number of circuit components increases. For this reason, it is often convenient to consider the entire membrane system as a "black box." A known signal is applied to the input and the resultant response is examined at the output as shown in Fig. 7.5. This information is then used to determine the nature of components in the black box.

Obviously, if we applied a constant voltage to the system some output would result, but this experiment produces very little information about the types and sizes of components within the system. To increase the total information available, it is more convenient to introduce a sinusoidal input at some fixed frequency and examine the response of the system for this frequency. If the components in the black box are linear—that is, capacitors and resistors and not diodes or transistors—the output of the circuit must remain at the same frequency. The "resistance" at this frequency $\omega = 2\pi v$ is called the impedance and is defined using Ohm's law,

$$V(\omega) = [Z(\omega)][i(\omega)] \qquad (7.32)$$

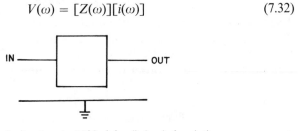

Fig. 7.5   An input–output "black box" circuit description.

The impedance defines the voltage response to the applied sinusoidal current at the frequency ω. We might expect to define the current response to an applied voltage in terms of this same parameter, but it is more convenient to define an admittance $Y(\omega)$, which produces an equation of the form

$$i(\omega) = [Y(\omega)][V(\omega)] \tag{7.33}$$

To determine the nature of $Z(\omega)$ and $Y(\omega)$, consider the response of the series and parallel RC circuits of Section 7.1 to sinusoidal inputs. For the series combination, the constant input voltage V is now replaced by

$$V(\omega) = V \sin(\omega t) \tag{7.34}$$

When Eq. (7.34) is substituted for V in Eq. (7.13), we must integrate an equation of the form

$$\int_0^t (d/dt)[q \exp(t/\tau)]\, dt = (V/R_m) \int_0^t [\exp(t/\tau) \sin(\omega t)]\, dt \tag{7.35}$$

and, because $q(0) = 0$,

$$q(t)[\exp(t/\tau)] = (V/R_m) \int_0^t [\exp(t/\tau) \sin(\omega t)]\, dt \tag{7.36}$$

The integral of a function of the form

$$\exp(ax) \sin(bx) \tag{7.37}$$

is

$$\exp(ax)[a \sin(bx) - b \cos(bx)]/(a^2 + b^2) \tag{7.38}$$

The right-hand side of Eq. (7.36) is

$$\exp(t/\tau)[\tau^{-1} \sin(\omega t) - \omega \cos(\omega t)]/(\tau^{-2} + \omega^2)|_0^t \tag{7.39}$$

$$\exp(t/\tau)[\tau^{-1} \sin(\omega t) - \omega \cos(\omega t) + 1]/(\tau^{-2} + \omega^2) \tag{7.40}$$

When both sides are multiplied by $\exp(-t/\tau)$, the expression for the current is

$$q(t) = (V/R_m)[\tau^{-1} \sin(\omega t) - \omega \cos(\omega t)]/(\tau^{-2} + \omega^2)$$
$$+ (V/R_m)[\exp(-t/\tau)]/(\tau^{-2} + \omega^2) \tag{7.41}$$

The second term has no oscillatory component and will rapidly decay to zero. The first terms will persist while the sinusoidal voltage is applied. The equation can be simplified to

$$q(t) = (V/R)\tau[\sin(\omega t) - (\omega\tau) \cos(\omega t)]/[1 + \omega^2\tau^2] \tag{7.42}$$

The sinusoidal voltage input has produced both a sine and cosine term. The combination can be reduced to a single sine function by introducing the following definitions:

$$\cos \delta = 1/(1 + \omega^2\tau^2)^{1/2} \qquad (7.43)$$

$$\sin \delta = \omega\tau/(1 + \omega^2\tau^2)^{1/2} \qquad (7.44)$$

The angle expresses a rotation produced by the product $\omega\tau$ relative to 1 as shown in Fig. 7.6. Equation (7.42) is now

$$VC[\sin(\omega t)\cos(\delta) - \cos(\omega t)\sin(\delta)]/(1 + \omega^2\tau^2) \qquad (7.45)$$

Using the identity

$$\sin(\omega t - \delta) = \sin(\omega t)\cos(\delta) - \cos(\omega t)\sin(\delta) \qquad (7.46)$$

the expression for charge reduces to a single sine term,

$$q(t) = (VC)(1 + \omega^2\tau^2)^{-1/2}\sin(\omega t - \delta) \qquad (7.47)$$

The phase delay δ can be determined from the ratio of Eqs. (7.44) and (7.43) or the geometry of Fig. 7.6 as

$$\tan \delta = \omega\tau \qquad (7.48)$$

so that the phase delay will be different for each frequency.

The phase delay reflects the fact that the charge on the capacitor cannot follow the amplitude of the applied voltage immediately. As the applied voltage peaks, this maximal applied potential will produce a current in the resistor, and the current requires a finite time to reach the capacitor. The charge on the capacitor will reach its maximum after the applied potential has reached its peak.

Equation (7.47) also indicates that the maximum charge attained by the capacitor is reduced as a function of frequency due to the term

$$(1 + \omega^2\tau^2)^{-1/2} \qquad (7.49)$$

At the higher frequencies, the capacitor has little time to charge fully before the polarity of the applied voltage reverses.

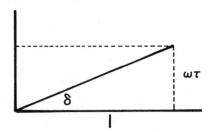

Fig. 7.6   The phase delay δ for a given ω.

The current to the capacitor plate is

$$i = dq/dt = (VC\omega)(1 + \omega^2\tau^2)^{-1/2} \cos(\omega t - \delta) \qquad (7.50)$$

while the potential buildup on the plate as a function of time is

$$V = q/C = V(1 + \omega^2\tau^2)^{-1/2} \sin(\omega t - \delta) \qquad (7.51)$$

The current and voltage at the capacitor are 90° out of phase, because the current must flow to the capacitor before its potential can build up.

If we wished to use the current and voltage of Eqs. (7.50) and (7.51) to define an impedance Z at this frequency, the ratio would be complicated by the presence of the sine and cosine terms. If we ignore these terms and consider only the leading terms, the impedance will be

$$Z = V/i = 1/\omega C \qquad (7.52)$$

We can include the 90° phase difference between current and voltage by introducing the imaginary j into the impedance expression, so that

$$Z = 1/j\omega C \qquad (7.53)$$

This definition suggests a simpler way to analyze circuits. The impedance Z is simply used as the "resistance" of the capacitor. This impedance will decrease with increasing frequency.

To illustrate this approach, the series RC combination is considered for a final examination. For an applied voltage $V = V \sin(\omega t)$, the potential equation is

$$V = i[R + (1/j\omega C)] \qquad (7.54)$$

where V and i are sinusoidal. The current in the circuit is then

$$i = V/[R + (1/j\omega C)] \qquad (7.55)$$

$$= j\omega CV/(1 + j\omega RC) \qquad (7.56)$$

Using the complex conjugate and $\tau = RC$, the equation can be resolved into a real and an imaginary part:

$$i = -\tau\omega^2 CV/(1 + \omega^2\tau^2) + j\omega CV/(1 + \omega^2\tau^2) \qquad (7.57)$$

These components are obviously 90° out of phase. The current expression can then be written completely if we introduce $\sin(\omega t)$ into the real term and $\cos(\omega t)$ into the imaginary term to give

$$i = (\omega CV)(1 + \omega^2\tau^2)^{-1} [\cos(\omega t) + \omega\tau \sin(\omega t)] \qquad (7.58)$$

This expression can also be generated by differentiating Eq. (7.42) for the charge.

The same approach can be used for parallel RC circuits described with the admittance Y. For a sinusoidal current, the relevant equation is

$$i = YV \tag{7.59}$$

where

$$Y = [(1/R) + j\omega C] \tag{7.60}$$

In such parallel circuits, the resistance is often replaced by the conductance,

$$G_m = 1/R_m \tag{7.61}$$

For an applied sinusoidal current, the voltage is

$$V = i/Y = i/(R_m^{-1} + j\omega C) \tag{7.62}$$

$$V = iR_m/(1 + j\omega\tau) \tag{7.63}$$

We shall return to this form in Section 7.4.

3. THE MEASUREMENT OF ELECTRICAL PARAMETERS

To determine both the resistance and capacitance of a membrane system at least two measurements are required. To develop two such measurements, the electrical response of the membrane can be measured when it is placed alternately in a series and a parallel configuration. For example, a constant voltage might be applied to a combination of a known resistance and the parallel $R_m C_m$ combination such as in Fig. 7.1. The capacitor of the membrane will charge with a time constant that must be dictated by both resistors, because they play a role in determining the total charge which reaches the capacitor. To illustrate this more clearly, consider the situation in Fig. 7.7 in which the capacitor has charged to its full potential. The resistor R_t is connected to the common ground, so that current can leave the capacitor through either R_t or R_m. The two resistors then constitute a parallel resistor combination, and the net resistance is

$$1/R = 1/R_t + 1/R_m \tag{7.64}$$

The time constant for this discharge is now

$$\tau = RC_m \tag{7.65}$$

and the potential will decay as

$$V = V_{max}[\exp(-t/\tau)] \tag{7.66}$$

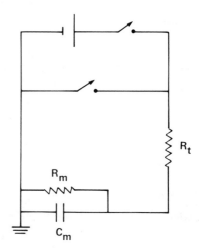

Fig. 7.7 A circuit to measure membrane resistance (R_m) and capacitance (C_m).

Additional information is still required to determine R_m and C_m from the time constant.

To understand the charging process, we need only realize that it is exactly the opposite of the discharging process. By applying a positive potential in place of ground and setting the potential between R_t and $C_m R_m$ to zero, the current would flow in exactly opposite directions with exactly the same time constant. However, the time constant is already known, so no new information is available with such a measurement.

Because R_t is chosen to be comparable in size to R_m, the applied potential will be divided between the two resistors at steady state. For the simple RC parallel circuit, the current at steady state was just

$$i = V/R_m \qquad (7.67)$$

because no current could flow through the capacitative branch when the capacitor was fully charged. For this system, the total current at steady state is

$$i = V/(R_t + R_m) \qquad (7.68)$$

and the potential drop across the test resistor is just

$$V_t = iR_t = V[R_t/(R_t + R_m)] \qquad (7.69)$$

Because R_t is known, R_m is determined from Eq. (7.69). The membrane capacitance can then be determined from Eq. (7.65).

An alternative approach that permits the determination of total capacitance and resistance for different oscillating frequencies involves matching the resistance and capacitance of the membrane with a known resistance and

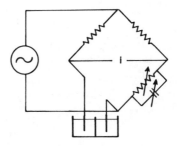

Fig. 7.8 An alternating current bridge.

capacitance using an alternating current bridge, as shown in Fig. 7.8. To understand the operation of the bridge, consider the balance point of the bridge when only resistance is present and all four resistors have the same value, as shown in Fig. 7.9. The bridge is connected to a voltage source as shown. Current will flow through each pair of resistors to ground with a current

$$i = V/2R \tag{7.70}$$

in each branch. The voltages at the midpoint of each branch are then equal,

$$V_{\mathrm{mpt}} = iR = V/2 \tag{7.71}$$

If the potentials on each side of the detector are equal, no current can flow in the detector branch. If the membrane resistor R_m differed from the value of the other three resistors, the potential at point A would differ from that at point B, and this difference would be detected. The alternating current bridge works on the same principle for each frequency studied. When the bridge is balanced, the R_m and C_m of the membrane must match the test R_t and C_t at that frequency.

   An alternating current bridge must be balanced for each frequency studied, so that the time scale for measuring with the device is quite long. It is more convenient to excite with all frequencies at once while observing the response

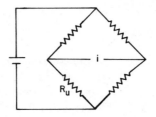

Fig. 7.9 Bridge balance when the bridge legs are exclusively resistive.

for these frequencies. For example, a noise voltage, containing a random distribution of frequencies, might be applied to the membrane. Instead of observing all frequencies, the noise introduces a random group of voltage frequencies that constitutes a cross section of all the frequencies. The distribution of frequencies in the applied noise potential and the current response must be examined in this case. This analysis can be simplified through the use of pseudo-random noise. A known distribution of frequencies that produces all the characteristics of random noise is applied to the system. Because the distribution of frequencies for the input is known, only the response must be analyzed. With such systems, a full impedance measurement over the full frequency range can be made in approximately 0.1 s.

4. COLE–COLE PLOTS

In Section 7.2, we determined that both the real and imaginary parts of the impedance depended on frequency and we developed an expression for admittance of a simple parallel RC circuit [Eq. (7.63)]. This expression can now be divided into its real and imaginary parts,

$$V = iR_m/(1 + j\omega\tau)$$

$$= iR_m/(1 + \omega^2\tau^2) - iR_m(j\omega\tau)/(1 + \omega^2\tau^2) \qquad (7.72)$$

$$V/i = R + j(-X) \qquad (7.73)$$

where R and $-X$ are the real and imaginary parts of the expression, respectively. In Fig. 7.10, R and $-X$ are plotted as a function of ω. The real resistance R will fall to zero at very high frequencies, because all the current is shunted through the capacitor. As ω approaches zero, all current is shunted

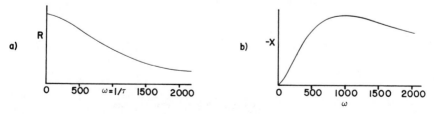

Fig. 7.10 (a) The real component of membrane impedance versus frequency ω. (b) The imaginary component of membrane impedance versus frequency ω.

through the resistor. The falloff of the graph toward zero is defined by a midpoint of $R_m/2$ when $\omega\tau = 1$, so that the time constant of the circuit matches the applied frequency.

The imaginary component plot rises to a maximum of $R_m/2$ when $\omega\tau = 1$. Thus, $R_m/2$ seems to play a central role in these expressions. The actual physical explanation for the form of the two curves will be explored in the next section. For the moment, we note that both components are equal when $\omega\tau = 1$.

The falloff with frequency is observed when an alternating potential is applied to suspensions of cells. A typical falloff for liver cells is shown in Fig. 7.11.

In place of the two distinct plots for R and $-X$, a combined plot, the Cole–Cole plot, can be used. In such plots, the real component is plotted on the abscissa while the imaginary component is plotted along the ordinate. The simple parallel circuit can be used to deduce the basic shape of the resultant curve in which each pair of R and X for a given ω are all plotted on a common graph.

To determine the shape of the graph, consider the high- and low-frequency limits and the case $\omega\tau = 1$. For $\omega = 0$, $R = R_m$ and $-X = 0$. At high frequency, $R = 0$ and $-X = 0$; the graph is bounded on the x-axis between 0 and R_m, and the values of $-X$ are zero at each extreme. At $\omega\tau = 1$, both R and $-X$ have the value $R_m/2$. The three points are shown in Fig. 7.12 as part of the total semicircular graph. The semicircular nature can be confirmed by using additional values of $\omega\tau$ and solving for R and $-X$.

The Cole–Cole plots can be used to describe the properties of a number of membrane systems. However, the experimental plots can be quite different from the plot developed from our simple RC circuit. For example, consider

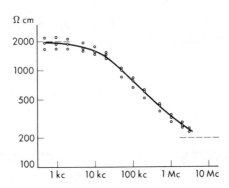

Fig. 7.11   The specific impedance of guinea pig liver in $\Omega \cdot$ cm as a function of the alternating current, measuring frequency. [Reproduced with kind permission of Cole, 1968].

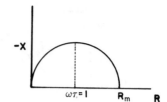

Fig. 7.12 A Cole–Cole plot for a membrane with R_m and C_m in parallel.

the effect of an additional resistance in series with the parallel circuit that produces Fig. 7.12. The expression is now

$$Z = R_s + R_m/(1 + \omega^2\tau^2) - jR_m\omega\tau/(1 + \omega^2\tau^2) \qquad (7.74)$$

The additional resistance is added to the real part of the impedance; every real component is now shifted along the real axis by an amount R_s, as shown in Fig. 7.13. The translation implies that the resistance R_s will still be present at infinite frequency.

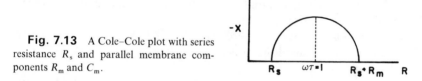

Fig. 7.13 A Cole–Cole plot with series resistance R_s and parallel membrane components R_m and C_m.

The RC electrical analog selected to develop the Cole–Cole plots had a single time constant. With biological membrane systems, it is possible to have more than one time constant. If two time constants are well separated, the Cole–Cole plot will produce two semicircular regions, as shown in Fig. 7.14.

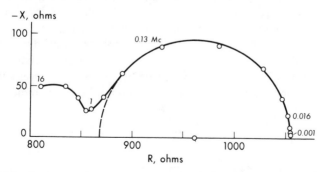

Fig. 7.14 Complex impedance locus for starfish (*Asterias*) egg suspension. [Reproduced with kind permission of Cole, 1968].

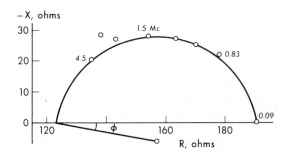

Fig. 7.15   Complex impedance plane representation for R versus $-X$ of the original data for calf red blood cell suspension. [Reproduced with the kind permission of Cole, 1968.]

When the system has a number of time constants that are relatively close together, an interesting phenomenon appears. We might expect that each time constant would give rise to its own Cole–Cole plot, so that the observed plot would be some complicated combination of the semicircles for each time constant. However, this combination of time constants produces a single semicircle. The major difference in this case is a depression of the center of the semicircle, as shown in Fig. 7.15.

The amount of semicircle depression can be expressed as an angle ϕ between the real axis and the radial line to R_s as shown.

5. COMPLEX DIELECTRIC CONSTANTS

Although the Cole–Cole plots can describe resistance and capacitative reactance, they can also be used to describe the dielectric constant behavior of the membrane as a function of frequency. To illustrate this phenomena, we consider an alternative development in which the real component of the capacitance is plotted on the real axis. The membrane has a dominant capacitative component, which is placed in series with a second resistive component R, which in this case, can be produced by resistive processes within.

The membrane current for an applied potential of maximum amplitude V is

$$i = V/[R + (j\omega C)^{-1}] \tag{7.75}$$

or

$$i = V(j\omega C)/(1 + j\omega RC) \tag{7.76}$$

$$= V(j\omega C)[1/(1 + \omega^2\tau^2) - j\omega RC/(1 + \omega^2\tau^2)] \tag{7.77}$$

The observed current can also be expressed in terms of the measured capacitance of the system,

$$q = CV \tag{7.78}$$

$$i = dq/dt = C\,dV/dt \tag{7.79}$$

For $V = V\exp(j\omega t)$, this expression becomes

$$i = j\omega CV\exp(j\omega t) \tag{7.80}$$

Equations (7.77) and (7.80) both represent current in the membrane system, so they must be equivalent. To generate both real and imaginary parts for Eq. (7.80), we note that the capacitance between the two electrodes would be C_0 in the absence of all solution and membrane between the electrodes. When the materials are reinserted, they introduce a relative dielectric constant so that

$$C = C_0\varepsilon \tag{7.81}$$

where the dielectric constant can be a complex function of the form

$$\varepsilon = \varepsilon' - j\varepsilon'' \tag{7.82}$$

Equating the two expressions gives

$$j\omega V\exp(j\omega t)[C_0(\varepsilon' - j\varepsilon'')]$$
$$= j\omega V\exp(j\omega t)[C/(1 + \omega^2\tau^2) - j\omega RC^2/(1 + \omega^2\tau^2)] \tag{7.83}$$

Equating real and imaginary terms from each side produces the following pair of relationships:

$$C_0\varepsilon' = C/(1 + \omega^2\tau^2) \tag{7.84}$$

$$C_0\varepsilon'' = \omega RC^2/(1 + \omega^2\tau^2) = C\omega\tau/(1 + \omega^2\tau^2) \tag{7.85}$$

The form for each of these equations is identical to the form that resulted from the study of the parallel RC circuit. However, the roles of the two expressions are now reversed. The real component describes the pure capacitative component, while the imaginary portion describes the resistance.

The dielectric-constant expressions now permit a physical interpretation of the real and imaginary components. The dielectric constant reflects the ability of the medium between the capacitor plates to reorient dipoles in the applied field. These aligned dipoles then act to "neutralize" some of the charge on the plates. For this reason, they enhance the charge storage capability of the capacitor, that is,

$$C = \varepsilon C_0 \tag{7.86}$$

and

$$q = CV = \varepsilon C_0 V \tag{7.87}$$

The real part of the dielectric constant decreases at the higher frequencies, suggesting that the dipoles have no time to realign with the field at these frequencies. The time constant τ now indicates the average time required for the dipoles to move in the field. For example, it would characterize the time required for the dipoles to resume their random orientations after an aligning electric field was removed.

In Chapter 2, we calculated the average dipole for dipoles in an electric field E. For an array of N dipoles, the total dipole moment is

$$P = \langle \mu \rangle N \tag{7.88}$$

where P, the polarization, is the total dipole moment produced by all the dielectric material in a unit volume. If a polarization P is produced by an applied electric field, then this polarization will decay in time if the electric field is removed. The relevant equation is

$$dP/dt = -P/\tau \qquad P(0) = P_0 \tag{7.89}$$

with a solution

$$P(t) = P_0 \exp(-t/\tau) \tag{7.90}$$

The net dipole moment decays to zero as the dipoles assume a random arrangement of configurations.

If the polarization is initially zero and an oscillating electric field

$$E = E_0 \exp(j\omega t)(\varepsilon_s - \varepsilon_\infty) \tag{7.91}$$

is applied, this field acts as a forcing function and the equation becomes

$$dP/dt = -P/\tau + E_0 \exp(j\omega t)(\varepsilon_s - \varepsilon_\infty) \tag{7.92}$$

Using the integrating factor $\exp(t/\tau)$ once again, the equation simplifies to

$$d/dt[P \exp(t/\tau)] = E_0 \exp(t/\tau + j\omega t)(\varepsilon_s - \varepsilon_\infty) \tag{7.93}$$

Integrating from 0 to t gives the solution

$$P = (\varepsilon_s - \varepsilon_\infty)[1/(1 + j\omega\tau)] \exp(j\omega t)[1 - \exp(-t/\tau)]E_0 \tag{7.94}$$

At a steady state,

$$P = (\varepsilon_s - \varepsilon_\infty)[1/(1 + j\omega\tau)] E_0 \exp(j\omega t) \tag{7.95}$$

so that the polarizability can also be resolved into real and imaginary components. Note that ε_∞ corresponds to the dielectric constant at high frequencies where molecular-motion dipole changes are not active, while ε_s corresponds to the real dielectric in the direct current limit. In order to resolve the dielectric into components, we note that the dielectric displacement, which is defined as the product of the dielectric constant and the electric field,

$$D = \varepsilon E \qquad (7.96)$$

is due to the sum of the field at infinite frequency plus the field generated by the aligned dipoles, that is, P. Thus,

$$D = \varepsilon_\infty E + P = \varepsilon E \qquad (7.97)$$

The total dielectric constant is then

$$\varepsilon E = \varepsilon_\infty E + (\varepsilon_s - \varepsilon_\infty)E/(1 + j\omega\tau) \qquad (7.98)$$

or $\qquad \varepsilon = \varepsilon_\infty + (\varepsilon_s - \varepsilon_\infty)/(1 + j\omega\tau) \qquad (7.99)$

which is now resolved into real and imaginary components

$$\varepsilon' = \varepsilon_\infty + (\varepsilon_s - \varepsilon_\infty)/(1 + \omega^2\tau^2) \qquad (7.100)$$

$$\varepsilon'' = (\varepsilon_s - \varepsilon_\infty)\omega\tau/(1 + \omega^2\tau^2) \qquad (7.101)$$

for comparison with Eqs. (7.84) and (7.85). Equation (7.99) includes ε_∞ to describe the capacitive component present in the absence of the polarizing material.

The derivation in terms of the polarizability emphasizes that the rotation of the dipole molecules involves a single time constant. For multiple time constants or a continuum of time constants, the Cole–Cole plot that results from a plot of ε' and ε'' will have a depressed semicircle center. How can Eq. (7.99) be modified to describe the effects of these more complex systems?

To reach any point on the nondepressed Cole–Cole plot, we moved horizontally to the point ε' and then moved vertically to ε''. Both terms contain the factor

$$(\varepsilon_s - \varepsilon_\infty)/(1 + \omega^2\tau^2) \qquad (7.102)$$

This indicates the following relationship between ε' and ε'':

$$\varepsilon'' = (\omega\tau)(\varepsilon' - \varepsilon_s) \qquad (7.103)$$

The two components are also at right angles, which may also be included by including j,

$$\varepsilon'' = (j\omega\tau)(\varepsilon' - \varepsilon_s) \qquad (7.104)$$

We can develop a similar relationship by considering any pair of chords u and v within the semicircle, as shown in Fig. 7.16. When the chords are inscribed in a semicircle, they are perpendicular to each other, and u and v are the hypoteneuses of the two right triangles A and B. If we determine u and v in this manner, we find

$$v/u = j\omega\tau \qquad (7.105)$$

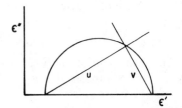

Fig. 7.16 The vectors u and v in a semi-circle that has its center on the ε' axis.

Our problem is now to determine the ratio v/u when the center of the semi-circle is depressed. When it is depressed all angles will be altered by a fractional parameter α, as shown in Fig. 7.17. The inscribed $90°$ angle increases to $90(1 + \alpha)$, while the angle between the two chords is decreased to $90(1 - \alpha)$. The inset of Fig. 7.17 shows the orientation of vectors ε' and ε'' if they are rotated until ε' intersects the semicircle. The depression of the semicircle then suggests that the angle between ε' and ε'' will decrease to $90(1 - \alpha)$.

Because

$$j = \exp[j(90^0)] = \cos(90°) + j\sin(90°) \qquad (7.106)$$

the depressed angle $90°(1 - \alpha)$ can be written as

$$\exp\{[j(90°)](1 - \alpha)\} = j^{1-\alpha} \qquad (7.107)$$

where $0 \le \alpha \le 1$. In all of these expressions j appears in conjunction with ω, and Cole and Cole noted that this requires a function of the form

$$\varepsilon = \varepsilon_\infty + (\varepsilon_s - \varepsilon_\infty)/[1 + (j\omega\tau)^{1-\alpha}] \qquad (7.108)$$

The ratio of v and u for the depressed Cole–Cole plot is

$$v/u = (\omega\tau)^{1-\alpha} \qquad (7.109)$$

Equation (7.109) does not provide absolute values for v and u, but it does show that the ratio is reduced for the depressed semicircles.

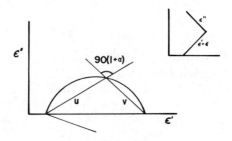

Fig. 7.17   The geometry of u and v for a semicircular plot with its center below the ε' axis.

Equation (7.108) provides a depression that is symmetrical about $\omega\tau = 1$. In some cases, a nonsymmetrical Cole–Cole plot is observed. The asymmetry can be described with an expression of the form

$$\varepsilon = \varepsilon_\infty + (\varepsilon_s - \varepsilon_\infty)/(1 + j\omega\tau)^\alpha \tag{7.110}$$

The introduction of $j^{1-\alpha}$ in Eq. (7.108) makes it difficult to resolve this expression into its real and imaginary components. Using

$$j^{1-\alpha} = \exp[j\pi(1 - \alpha)/2] \tag{7.111}$$

the resulting components are

$$\varepsilon' = \varepsilon_\infty + (\varepsilon_s - \varepsilon_\infty)\frac{1 + (\omega\tau)^{1-\alpha}\sin(\pi\alpha/2)}{1 + (\omega\tau)^{2(1-\alpha)} + 2(\omega\tau)^{1-\alpha}\sin(\pi\alpha/2)} \tag{7.112}$$

$$\varepsilon'' = (\varepsilon_s - \varepsilon_\infty)\frac{(\omega\tau)^{1-\alpha}\cos(\pi\alpha/2)}{1 + (\omega\tau)^{2(1-\alpha)} + 2(\omega\tau)^{1-\alpha}\sin(\pi\alpha/2)} \tag{7.113}$$

Despite the complexity of these relations, the average value of τ can always be found from Eq. (7.109). For a given frequency, the chords u and v are drawn and measured. The angle α is determined at the intersection of the chords, and τ is determined from these parameters.

The center of the semicircle is located at the point,

$$(\text{Re, Im}) = [\tfrac{1}{2}(\varepsilon_s + \varepsilon_\infty), -\tfrac{1}{2}(\varepsilon_s - \varepsilon_\infty)\tan(\pi\alpha/2)] \tag{7.114}$$

6. THE DEBYE RELAXATION TIME

The real portion of the complex dielectric constant decayed at higher frequencies, and this was attributed to the inability of the molecules of the dielectric to respond to the changes in electric field. The time constant τ defined both the midpoint of this fall and the point of maximal energy dissipation, as expressed by the imaginary component of the dielectric constant. Thus, the time constant can be related to the motions of the molecules of the dielectric. If the dielectric medium contains molecules with permanent dipole moments, then the time constant must be related to the time required for them to rotate under the influence of an applied electric field. This relationship will be developed to express the time constant of the system as a function of the molecular parameters of the individual dipoles.

When a dipole is placed in an electric field, it experiences a torque that works to align the dipole with the field. In the absence of all other effects, all particles would align with the field. This alignment process is frustrated by thermal interactions between the molecules or other system molecules acting to randomize the orientation of the dipole. The competition between two effects at a constant applied field led to the Langevin expression for an average dipole moment. However, the Langevin expression gave no data on the time that each dipole spent in each orientation; it merely described the average number to be found at each orientation at any time.

In order to introduce time into the dipole rotation process, we must introduce a frictional component. The medium around each dipole is assumed to be a continuous fluid. As the fluid viscosity increases, the dipole will rotate more slowly in the applied field.

The electric field is related to force, and forces produce acceleration. Thus, the dipoles might be expected to undergo a rotational acceleration when a field is applied. However, the fluid exerts a frictional force directly proportional to the velocity.

$$F_f = \zeta \, d\theta/dt \tag{7.115}$$

where ζ is the angular frictional coefficient. Because of friction, the rotating dipole will accelerate until the angular velocity reaches the point at which the frictional force just balances the applied torque. The dipole reaches this point rapidly and then continues to rotate at a constant velocity.

The frictional force can be related to the randomizing effects of thermal agitation and the aligning force of an applied electric field with the Debye equation for rotational diffusion,

$$\zeta \, \partial f/\partial t = (1/\sin \theta) \frac{\partial}{\partial \theta} \left\{ (\sin \theta) \left[kT \left(\frac{\partial f}{\partial \theta} \right) - \mu E (\sin \theta) f \right] \right\} \tag{7.116}$$

where $f(\theta, t)$ is the fraction of dipoles in the angular region between θ and $\theta + d\theta$ at time t. Equation (7.116) can be used to determine a rotational relaxation time without solving the equation in detail.

The rotational diffusion equation can be simplified if we consider the response of the dipoles when the electric field is suddenly set to zero. For $E = 0$, the equation is

$$\partial f(\theta, t)/\partial t = (1/\sin \theta) \frac{\partial}{\partial \theta} \left[(\sin \theta)(kT/\zeta) \frac{\partial f}{\partial \theta} \right] \tag{7.117}$$

Even though the field has been removed, the fraction $f(\theta, 0)$ is determined by the field applied until time zero. The fractions of dipoles at each value of θ are then proportional to the Boltzmann factor,

$$f(\theta, 0) = \exp(+\mu E \cos \theta/kT) \tag{7.118}$$

Because the field is removed at $t = 0$, this distribution should decay to a random distribution, such that

$$\exp(0) = 1 \qquad \text{for all } \theta \qquad (7.119)$$

In order to reach this equiprobable distribution, a time-dependent function $\phi(t)$ is introduced into Eq. (7.120) to give the trial solution

$$f(\theta, t) = \exp\{\mu E[\phi(t)] \cos \theta / kT\} \qquad (7.120)$$

As $\phi(t)$ decays to zero, f will approach the equiprobable distribution.

The exponential can be expanded to give a simpler trial function for f,

$$f = 1 + (\mu E/kT)[\phi(t)] \cos \theta \qquad (7.121)$$

This trial function is now substituted into Eq. (7.117),

$$(\mu E/kT)(\cos \theta)\partial\phi/\partial t = (1/\sin \theta)(\partial/\partial\theta)\{(\sin \theta)(kT/\zeta)(-\sin \theta)[\phi(t)]\} \qquad (7.122)$$

$$= (1/\sin \theta)(-2 \sin \theta \cos \theta)(kT/\zeta)(\mu E/kT)[\phi(t)] \qquad (7.123)$$

Cancelling common terms gives

$$d\phi(t)/dt = (2kT/\zeta)[\phi(t)] = -\phi(t)/\tau \qquad (7.124)$$

The decay of the dipole distribution is described by a first-order equation with a time constant

$$\tau = \zeta/2kT \qquad (7.125)$$

The frictional coefficient for a sphere is dictated by the viscosity η of the fluid in which the rotation takes place and the radius of the sphere via Stokes's relation,

$$\zeta = 8\pi\eta a^3 \qquad (7.126)$$

where a is the sphere radius. If Eq. (7.126) is introduced for the frictional coefficient, the relaxation time is

$$\tau = 4\pi\eta a^3/kT \qquad (7.127)$$

The Debye equation gives time constants approximately 5 to 10 times higher than the experimentally observed relaxation times. The discrepancy is related to the description of the surrounding medium as a continuum. To compensate, it is necessary to include the size of the surrounding molecules. If these molecules have a radius a', then the expression for the time constant is

$$\tau = 4\pi\eta a^3(a/6a')/kT \qquad (7.128)$$

7. THE CAPACITANCE FOR SUSPENSIONS OF CELLS

In our studies of membrane capacitance, we have assumed that a membrane, such as a planar bilayer, is placed between two electrodes. The system impedance is then measured with a bridge. Interestingly, the effects of membrane capacitance can also be observed when a suspension of cells in a conducting saline solution is placed between the electrodes. Current will flow in the solution, but it will be affected by the presence of the cells in the medium. The perturbing effect of cells in the medium can be measured to deduce the membrane capacitance for these cells.

Consider the situation illustrated in Fig. 7.18, in which a dielectric sphere is inserted into a homogeneous electric field. Current cannot flow through the sphere because the dielectric is an insulator. However, dipolar molecules of the dielectric will align with the field. The aligned dipoles will produce a field that will perturb the homogeneous applied field near the dielectric. We must determine this perturbed field and any potential differences that might arise as a result of this perturbation.

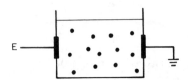

Fig. 7.18 Nonconducting spheres (cells) in an applied electric field E.

Because there are no free charges in the region, Poisson's equation will reduce to the Laplace equation. This equation is written in polar coordinates as

$$(1/r^2)\frac{\partial}{\partial r}(r^2 \partial V/\partial r) + [1/(r^2 \sin \theta)](\partial/\partial \theta)(\sin \theta \, \partial V/\partial \theta) = 0 \quad (7.129)$$

for the potential at any point in the space of interest. The origin of the coordinate system is chosen at the center of the sphere, while the angle θ is defined as $0°$ along the direction of the field, that is, the z direction. The coordinate system is shown in Fig. 7.19.

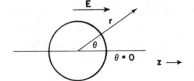

Fig. 7.19 Coordinates and orientation for a single sphere in an applied electric field. The field is directed only along the z axis.

Although Laplace's equation appears complex, it possesses a standard solution,

$$V = Ar \cos \theta + \frac{B \cos \theta}{r^2} \qquad (7.130)$$

which can be checked by inserting it into Eq. (7.129) and performing the appropriate operations.

Although Eq. (7.130) is a valid solution for Laplace's equation, it does not yet describe the potential near the dielectric sphere. To determine the full solution, appropriate values for the constants A and B must be determined. These values are determined by using certain boundary conditions that define the potential at certain positions in space, that is, (r, θ). If the constants provide the proper potential at some known location, they are expected to provide the proper potential at all other spatial locations as well. Because there are two constants, two such boundary conditions are required.

A single sphere inserted into the homogeneous electric field should have little effect on this field in regions far removed from the sphere. Thus, as r goes to infinity, the potential should be defined by the homogeneous electric field

$$E = -\partial V / \partial r \qquad \text{for} \quad r \to \infty \qquad (7.131)$$

Figure 7.20 shows two cases of interest. If r is selected perpendicular to the field direction, the lines of force continue linearly without experiencing a perturbation. If the vector r places us sufficiently far in front of the sphere, the electric field in this region has not yet experienced the sphere. The electric field behind the sphere is particularly interesting, because it appears that the lines of force have passed the sphere and should be distorted by it. However, because there is no net charge on the sphere, the electric field leaving a cylinder that encompasses the sphere will leave that cylinder unaltered. The electric field is homogeneous in all directions a large distance from the sphere.

To use this first boundary condition, the potential expression must be differentiated with respect to r,

$$E = -\partial V / \partial r = -A \cos \theta + (2B \cos \theta)/r^3 \qquad \text{as} \quad r \to \infty \qquad (7.132)$$

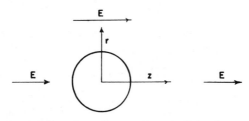

Fig. 7.20 The electric field at a large distance r from the dielectric, nonconducting sphere.

The second term has an inverse cube dependence, which reduces it to zero as r approaches infinity. Thus,

$$E = -A \cos \theta \qquad (7.133)$$

This condition must hold at all angles, including $\theta = 0$, and this requires that

$$E = -A$$

Thus, for any value of r and θ,

$$E = E \cos \theta - (2B \cos \theta)/r^3 \qquad (7.134)$$

and

$$V = -Er \cos \theta + (B \cos \theta)/r^2 \qquad (7.135)$$

The $r \cos \theta$ in the first term is necessary because work is done only along the field direction. If we move along a radial vector perpendicular to the electric field, no work is done. The condition $\cos \theta = 90°$ specifies no component along the electric field. If a charge is moved from r_1 to r_2 parallel to the field, the potential difference will be

$$V = -Er_2 \cos(0°) - [-Er_1 \cos(0°)] = -E(r_2 - r_1) \qquad (7.136)$$

All these calculations are performed at large r, so that the second term in the potential expression can be ignored.

In order to define B for the second term, a second boundary condition is necessary. Because the dielectric sphere is not a conductor, the electric field lines cannot enter the sphere. The field must be continuous, so it must drop to zero as it approaches the surface of the sphere at any point. This will cause the lines of force to curve around the sphere. To express this boundary condition, we state that E must be zero for every $r = a$, that is,

$$\partial V/\partial r = 0 \qquad \text{when} \quad r = a \qquad (7.137)$$

or, using Eq. (7.134),

$$Q = E \cos \theta - (2B \cos \theta)/a^3 \qquad (7.138)$$

Solving for B gives

$$B = (-Ea^3 \cos \theta)/(2 \cos \theta) = -Ea^3/2 \qquad (7.139)$$

When this constant is substituted into the electric field expression, this electric field can be determined for any value $r \geq a$,

$$E = -E \cos \theta - E(a/r)^3 \cos \theta \qquad (7.140)$$

The potential for any value of r is

$$V = -Er \cos \theta - (Ea^3 \cos \theta)/2r^2 \qquad (7.141)$$

As expected, the electric field in Eq. (7.140) does fall to zero at $r = a$. However, the potential at the surface is not zero! At $r = a$,

$$V = -Ea \cos \theta - (Ea^3 \cos \theta)/2a^2$$
$$= (-3Ea \cos \theta)/2 \qquad (7.142)$$

This potential represents the energy necessary to bend the lines of force around the sphere. The potential is associated with the motion of a test charge, so this description is equivalent to stating that we have "charged" the sphere to deflect a test charge moving along the lines of force.

The lines of force can be described more clearly when we note (see Chapter 10) that Ohm's law can be written in the form

$$I = \kappa E \qquad (7.143)$$

where I is the current through a unit area perpendicular to the field E and κ is the conductivity of the medium. Because of the direct proportionality, the current flow will follow the electric lines of force. Moving charges that constitute the current curve around the nonconducting sphere are shown in the upper left frame of Fig. 7.21.

Because a charge appears on the dielectric sphere when it is placed in the electric field, each sphere becomes a charge storage device, that is, a capacitor. The $\cos \theta$ term defines the distribution of the stored charge on the surface. When $\cos \theta = \cos 0° = +1$, the potential is

$$V = -3Ea/2 \qquad (7.144)$$

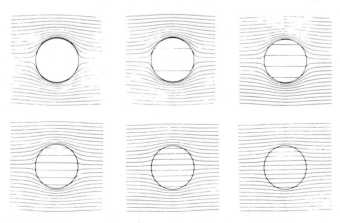

Fig. 7.21 Lines of current flow around spheres placed in a uniform field. The approximate ratios of the conductances of sphere to medium, from left to right were [top] 0, 0.02, 0.08, and [bottom] 0.18, 0.35, 0.6. [Reproduced with kind permission of Cole, 1968.]

On the left side of the sphere where $\cos(180°) = -1$, the potential is

$$V = +3Ea/2 \tag{7.145}$$

The sphere has become a dipole. The plus and minus charge is not concentrated at two separate locations as it would be for a simple dipole, because the $\cos\theta$ term distributes the charge over the entire surface.

Because each spherical dipole contributes to the total capacitance of the solution, it should now be possible to calculate this total capacitance. If a charge dq is transferred to the surface, an amount of energy

$$dU = Vdq \tag{7.146}$$

is required. However, dq also depends on both the sphere capacity C and the voltage on this sphere,

$$dq = d(CV) = C\,dV \tag{7.147}$$

The energy is then

$$dU = CV\,dV \tag{7.148}$$

The total energy stored in the capacitor can then be found by integrating this expression from the uncharged sphere ($V = 0$) to the final potential produced by an electric field. This gives

$$U = \int_0^V CV\,dV = \tfrac{1}{2}CV^2 \tag{7.149}$$

for the simple case where the capacitance is a constant and V does not vary with position on the sphere.

Because the potential does vary with its location on the sphere, we must integrate the stored potential at every point on the membrane surface. If c_s is the capacitance per unit area, then the energy expression must be modified to

$$U = \tfrac{1}{2}\int_0^V c_s V^2\,dS \tag{7.150}$$

A volume element on the sphere of radius a is

$$dS = a^2 \sin\theta\,d\theta\,d\phi \tag{7.151}$$

The variable ϕ, which describes changes around the equator of the sphere ($\theta = 90°$), does not appear in the potential expression. It can be integrated from 0 to 2π to give

$$dS = 2\pi a^2 \sin\theta\,d\theta \tag{7.152}$$

Equation (7.142) for the potential and Eq. (7.152) are now substituted into Eq. (7.150) to give

$$U = \tfrac{1}{2} \int_0^\pi c_s[(3a^3 \cos \theta) E/2]^2 (2\pi a^2) \sin \theta \, d\theta$$

$$= (9c_s \pi a^4 E^2/4) \int_0^\pi \cos^2 \theta \sin \theta \, d\theta \qquad (7.153)$$

The integral becomes

$$\left. \frac{-\cos^3}{3} \right|_0^\pi = \frac{2}{3} \qquad (7.154)$$

and the final expression is

$$U = (3c_s \pi a^4 E^2/2) \qquad (7.155)$$

If there are n spheres in some unit volume of solution, the total energy is

$$U_t = nU = nc_s(3a^4 \pi E^2/2) \qquad (7.156)$$

If this unit volume were observed, it would appear to have some total capacitance C. From this point of view, the total energy per unit volume is

$$U_t = CV^2/2 \qquad (7.157)$$

Because the energies must be identical, they can be equated to find an expression for c_s in terms of the measured capacitance C of the solution, the electric field, and the density and radius of the spheres:

$$CV^2/2 = nc_s(3a^4 \pi E^2/2) \qquad (7.158)$$

or

$$c_s = CV^2/3\pi a^4 nE^2 \qquad (7.159)$$

Because we are dealing with a unit volume of solution, the potential and electric field are identical since the field is just potential per unit length. Thus, the equation for the capacitance of a unit area of membrane is just

$$c_s = C/3\pi a^4 n \qquad (7.160)$$

The total volume of all n spheres in the unit volume is

$$v_s = (4/3)\pi na^3 \qquad (7.161)$$

This volume can be determined using conductivity measurements. When it is inserted into Eq. (7.160), we find

$$c_s = 4C/9av_s \qquad (7.162)$$

Using Eq. (7.162) for a solution with a ratio of $v_s/v_t = 0.21$ for red blood cells, Fricke determined a total capacitance of 129 pF. The cells had an equivalent radius of 2.6 μm to produce a unit membrane capacitance of 1.0 $\mu F/cm^2$. This magnitude is similar to that observed for bilayer membranes measured by the more direct methods discussed at the beginning of the chapter.

8. MEMBRANE CAPACITANCE FOR CONDUCTING CELLS

In the previous section, we chose the simple example of nonconducting spheres so that the electric field became zero at the membrane surface. If the membrane does have a finite conductance, some current or lines of force can pass through the surface. In such a case, we must calculate the electric fields for both the interior and exterior regions near the cell. This requires modification of our boundary conditions.

Because no extra charge has been added, the Laplace equation is again valid. However, the constant terms in the solution must be evaluated for both the interior and exterior regions. Thus, four constants must be evaluated for two potential functions that must match at the cell surface,

$$V_e = (Ar \cos \theta) + (B \cos \theta)/r^2 \tag{7.163}$$

$$V_i = (Cr \cos \theta) + (D \cos \theta)/r^2 \tag{7.164}$$

Only the external solution extends to large r. Using the boundary condition

$$-\partial V_e/\partial r = E \qquad \text{as} \quad r \to \infty \tag{7.165}$$

we again find

$$E = -A \tag{7.166}$$

In the interior region, the potential must be finite everywhere, including $r = 0$. The second term in Eq. (7.164) would become infinite at $r = 0$, and this indicates that this term cannot be valid in the interior region—that is, D must be zero. The two equations are now

$$V_e = (-Er \cos \theta) + (B \cos \theta)/r^2 \tag{7.167}$$

$$V_i = Cr \cos \theta \tag{7.168}$$

To evaluate the two remaining constants, two additional boundary conditions are required for the membrane surface. The variations in potential and electric field at the surface must be determined.

At the surface, the potential must be continuous, so that

$$V_e = V_i \qquad \text{at} \quad r = a \tag{7.169}$$

Although the electric field need not be continuous at the boundary, the net current across any boundary must be. The current density was defined using Ohm's law,

$$I = \kappa E \qquad (7.170)$$

The equivalent conductance of a unit volume of solution is defined as Λ, so that the current in the external medium is

$$I_e = \Lambda_e E_e \qquad (7.171)$$

while

$$I_i = \Lambda_i E_i \qquad (7.172)$$

For continuity of current,

$$\Lambda_e E_e = \Lambda_i E_i \qquad (7.173)$$

for fields defined in the radial direction, that is,

$$\Lambda_e \, \partial V/\partial r|_{r=a} = \Lambda_i \, \partial V/\partial r\,|_{r=a} \qquad (7.174)$$

Substituting Eqs. (7.167) and (7.168) into this equation gives

$$\Lambda_e[(-E \cos \theta) - (2B \cos \theta)/a^3] = \Lambda_i(C \cos \theta) \qquad (7.175)$$

Equation (7.169) for the equality of potentials at the boundary gives

$$(-Ea \cos \theta) + (B \cos \theta)/a^2 = Ca \cos \theta \qquad (7.176)$$

These expressions produce two simultaneous equations in the two unknown constants, B and C,

$$\Lambda_e E + 2\Lambda_e B/a^3 = -\Lambda_i C \qquad (7.177)$$

and

$$-E + B/a^3 = C \qquad (7.178)$$

The lower equation can be multiplied by Λ_i and added to the first to give

$$(\Lambda_e - \Lambda_i)E = -(B/a^3)(\Lambda_i + 2\Lambda_e) \qquad (7.179)$$

or

$$B = -(\Lambda_e - \Lambda_i)a^3 E/(\Lambda_i + 2\Lambda_e) \qquad (7.180)$$

$$= -a^3 E[1 - (\Lambda_i/\Lambda_e)]/[2 + (\Lambda_i/\Lambda_e)] \qquad (7.181)$$

The constant C is found by multiplying the lower equation by $2\Lambda_e$ and subtracting to give

$$C = -3\Lambda_e E/(2\Lambda_e + \Lambda_i) \qquad (7.182)$$

The potential equations are

$$V_e = -Er \cos \theta - \{(a^3 E/r^2)[1 - (\Lambda_i/\Lambda_e)]/[2 + (\Lambda_i/\Lambda_e)]\} \cos \theta \quad (7.183)$$

and

$$V_i = (-3Er \cos \theta)[\Lambda_e/(2\Lambda_e + \Lambda_i)] \quad (7.184)$$

At the boundary where $r = a$,

$$V = (-3Ea\Lambda_e \cos \theta)/(2\Lambda_e + \Lambda_i) \quad (7.185)$$

If the spheres become nonconducting ($\Lambda_i = 0$), the equation reduces to our previous result,

$$V = (-3Ea \cos \theta)/2 \quad (7.186)$$

As the conductivity of the sphere increases, the potential maintained at the surface of the sphere will decrease. When $\Lambda_e = \Lambda_i$, the equation reduces to

$$V = -Ea \cos \theta \quad (7.187)$$

The electric field is homogeneous throughout the medium.

Lines of force for different conductivities are illustrated in Fig. 7.21.

PROBLEMS

1. An equivalent circuit often used to model the electrical properties of a membrane is

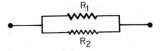

(a) Calculate the real and imaginary components of the impedance. (b) Calculate $|Z|$, the absolute value of the impedance, and ϕ, the phase shift, associated with this circuit.

2. Consider the resistive–inductive circuit,

If the voltage across the inductor is defined by $V = L \, di/dt$, determine a relation between the current and voltage across the inductor for some frequency ω.

3. Prove that

$$Z = [R + j(-X)] = [R_\infty + (R_0 - R_\infty)/(1 + \omega^2\tau^2)] - j[\omega\tau(R_0 - R_\infty)/(1 + \omega^2\tau^2)]$$

produces a semicircle in the $(R, -X)$ plane.

4. A sphere immersed in a suspension of conductivity Λ_1 consists of two regions: an inner core of conductivity $\Lambda = 0$ and radius a, and an outer core of radius b and conductivity Λ_2. Set up the general solution for this equation and find the boundary conditions for each boundary.

5. The maximal value of $\varepsilon'' = D\varepsilon_0$ in a Cole–Cole plot corresponds to $D = 8$. At low frequencies, the relative dielectric constant is 40. It drops to 20 at high frequency. (a) Is this a leaky on nonleaky capacitor? Why? (b) Determine the parameter α if this semicircle is depressed.

6. The following relative dielectric constants are observed for a membrane at different frequencies

Frequency	ε_r'	ε_r''
10 Hz	5	0.1
100 Hz	5	0.2
1 kHz	3	0.2
10 kHz	1	0.2
100 kHz	1	0.1

(a) What are ε_∞ and ε_0 for this system? (b) What is the relaxation time for the molecules?

7. Show that the imaginary component for a depressed Cole–Cole plot is smaller than the imaginary component for the undepressed Cole–Cole plot.

8. (a) Calculate the potential at the surface of a nonconducting sphere of radius 1 μm in a constant field of 1000 V/m. (b) When the sphere is conducting, determine the potential at the sphere surface for 1000 V/m when $\Lambda_e = 150 \text{ S} \cdot \text{cm}^2/\text{equiv}$ and $\Lambda_i = 100 \text{ S} \cdot \text{cm}^2/\text{equiv}$.

9. A nerve axon can be approximated as a long cylinder. Assume such a cylinder is immersed in a solution between two plates that support a field E in the solution. The field is perpendicular to the cylinder surface. (a) Is this a two- or three-dimensional problem? Why? Which variables must be used in a solution of Poisson's equation? (b) In the analysis for spheres and cylinders in an electric field, the energy is proportional to $\cos^2 \theta$. For a cylinder of length b and radius a, determine a factor that will give the average potential on the cylinder.

Chapter 8

The Electrical Double Layer

1. THE HELMHOLTZ DOUBLE LAYER

The potential across a membrane can be measured by placing electrodes in the two bathing solutions and recording a potential difference. Such measurements do not reveal where the energy was dissipated as a test charge moved through the membrane. For the pH electrode discussed in Chapter 4, the thermodynamic potential was produced between the external solution and sites within the glass electrode. In a biological membrane, the dissipation of the transmembrane potential can also be affected by a charged surface. Figure 8.1 describes a simplified membrane with a single negatively charged surface. A positive potential is applied across this membrane as shown. If there were no negative surface charge, a test particle would approach the surface and its potential would dissipate steadily as it crossed the homogeneous membrane, as shown by the dotted line in the figure. With the negative charge present, the surface would have a lower potential than the solution and the test charge would move to this lower potential. In order to escape from the surface, it would now have to acquire the energy to separate from the negative charge and enter the membrane. If a collision provides sufficient energy to separate it from the membrane charge, it will then have a reasonable probability of crossing the membrane. If the left side of the membrane is at a lower potential than the right side, the particle must ultimately reach that side, even though the details of its journey are not known. The time required for such a journey, however, can be a very sensitive function of the microscopic structure of the membrane.

Our hypothetical membrane with its layer of negative charge raises an interesting point. Most of the phases that we have considered in earlier

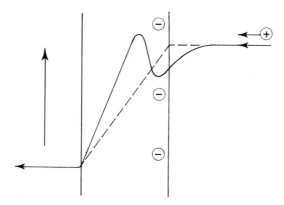

Fig. 8.1 Potential dissipation for a test charge moving through a membrane, where the dotted line is linear potential drop for a homogeneous membrane with no embedded charge, and the solid line is potential charge for a positive test charge approaching a membrane with imbedded negative surface charge.

chapters have been electroneutral; electroneutrality is also expect d here. Consider a model system in which the membrane can be formed before the bathing solutions are added. This membrane phase must be electroneutral, and each of the negative charge sites will be balanced by a cation, such as Na^+. When the bathing solutions are added, however, the cations will ionize. If the volume of the solutions is sufficiently large, the thermal motions of the solutions will disperse the ions, leaving the negative surface discussed earlier. This, however, is an extreme case. After ionization, a negatively charged surface remains and should exert a coulombic attraction on the cations. It is the dynamic competition between the surface and the aqueous phase that produces the phenomena to be studied.

Once the cations leave the surface, they can return in two distinct ways. In the first case, they can shed their water of hydration and rebind to the negative charge sites. This might be expected to simply neutralize the charge site, although we shall see that this recombination can be more complicated. In the second case, the cation is attracted to the surface but the attraction is insufficient to strip the cation of its waters of hydration. The cation will be attracted toward the surface but will be stopped at a distance equal to the radius of its hydration sheath. The two possibilities are described in Fig. 8.2. Note that the ions that approach the surface need not be the same ions that originally belonged to the membrane. If other cations are present in solution, these cations might be attracted more strongly. These solution cations will be balanced by an equal number of anions, so that overall electroneutrality is maintained.

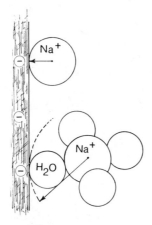

Fig. 8.2 Closest approach of hydrated and unhydrated cations to a negatively charged surface.

Helmholtz postulated that all the cations necessary to neutralize the surface charge formed a layer of positive charge at some fixed distance from the membrane surface, for example, the distance of closest approach of the hydrated ions. For a planar surface, there would now be two planar layers of charge of opposite sign separated by some constant distance. This is equivalent to a parallel-plate capacitor where the charge on the capacitor is determined by the number of charges on the surface. Thus, if the capacitance C can be determined from the geometry of the surfaces, the potential between the positive and negative plates can be determined from the basic definition of capacitance as the charge stored per given voltage,

$$C = q/V \qquad (8.1)$$

or

$$V = q/C \qquad (8.2)$$

where q is the charge in coulombs while V is the potential difference. For a parallel-plate capacitor, the capacity will increase as the area of the plates increases and as the separation between the plates decreases. In addition, as noted in Chapter 3, the dielectric medium between the plates will tend to "neutralize" some of the charge on the plates so that a larger dielectric constant for this medium will permit a large capacity. This information can be compressed into the equation for a parallel-plate capacitor,

$$C = \varepsilon_r \varepsilon_0 A/d \qquad (8.3)$$

where ε_0 is the permittivity constant (8.85×10^{-12} F/m), ε_r is the relative dielectric constant of the medium, d is the plate separation, and A is the area.

Because this capacitance is often expressed in cgs units for double layer studies, the expression in these units is provided for reference,

$$C = \varepsilon_r A/4\pi d \tag{8.4}$$

Note that the MKS expression was simply divided by $4\pi\varepsilon_0$.

Equations (8.2) and (8.3) can now be combined to produce an expression for the potential in terms of the physical properties of the double layer,

$$V = q(d/\varepsilon_r\varepsilon_0 A) = (d/\varepsilon_r\varepsilon_0)(q/A) = [d/\varepsilon_r\varepsilon_0]\sigma \tag{8.5}$$

The charge per unit area, $\sigma = q/A$, indicates that it is this charge "density" rather than the absolute charge that determines the potential. In cgs units, the Helmholtz double layer potential is

$$V = (4\pi d/\varepsilon_r)\sigma \tag{8.6}$$

The cgs units for σ must be esu/cm^2.

The potential experienced by a charged positive test particle approaching the Helmholtz double layer is diagrammed in Fig. 8.3. As the particle approaches the positive charge layer, it is repelled and its potential increases. Once the charge reaches the intralayer region, it moves from the repulsive positive plate toward the negative charge plate. Because the medium between the two plates is assumed homogeneous, the electric field experienced by the test charge in this region is constant and the potential falls linearly, as shown in the figure. The net effect of crossing the double layer is a potential drop of V. Note, however, that we have calculated the potential difference and not the absolute potentials on the plates.

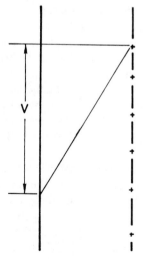

Fig. 8.3 Potential dissipation in a Helmholtz double layer.

2. THE GOUY–CHAPMAN THEORY

Although the Helmholtz model provides a good initial estimate for the potential at an interface, a number of rather drastic assumptions are made in the development of the model. For example, the anions are assumed to form a homogeneous sheet of charge, and the cations in solution form a complementary sheet unaffected by the thermal motions of the solution. Obviously, as the temperature of the solution is increased, these motions must exert a disruptive effect on the Helmholtz sheet of charge.

The Gouy–Chapman model permits the "sheet" of charge in solution to be a diffuse mixture of anions and cations. This sheet will not be confined to a plane but will occupy a finite volume of solution that contains both cations and anions. There will be excess cations to balance the negative charges on the membrane surface. To describe this volume of charge, it is divided into a series of planar regions. The planar sheet closest to the membrane surface will have the largest excess of cations over anions, although there will be insufficient cations in this plane to balance the total surface charge. The next sheet will have a smaller excess of cations. At large distances, the number of anions and cations will become equal, that is, the solution will be homogeneous. If we sum the excess cations in each of the planes, the total charge will just balance the surface charge. The arrangement is shown in Fig. 8.4 for a surface with 10 negative surface charges per unit area. For this simple model, the innermost plane contains 6 net positive charges, the second plane contains 3, and the outermost unbalanced plane contains 1. The innermost plane actually contains 9 cations and 3 anions; it is the excess of 6 cations in this region that is important.

The excess ions found at various distances from the surface will depend on the number of charges on the membrane surfaces. In addition, the number will depend on the total number of ions in the solution. If a large number of such ions are available, it becomes more likely that random thermal motions will bring ions near the surface. However, because the surface is a layer of negative charge, cations are expected to be attracted into this region while anions will tend to be repelled from the region. Both effects will contribute to the cationic excess.

The negative charge on the surface will produce a potential in the solution. However, as cations are attracted toward the surface by this potential, they will cancel some of this negative potential. As the distance from the surface increases, more of the surface-generated potential will be cancelled or "screened" by the cations. Thus, the potential that is observed at any distance from the surface will be a function of both the surface charge and the total concentration of excess ions closer to the surface.

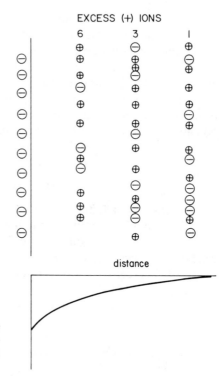

EXCESS (+) IONS

Fig. 8.4   Distribution of solution charge near a negatively charged surface. The negative surface charge is balanced by an equal number of extra cations in solution.

The potential V and the charge are related through Poisson's law (Chapter 3), which has the form

$$\nabla^2 V = -\rho(x, y, z)/\varepsilon_0\varepsilon_r \qquad (8.7)$$

in three dimensions. However, most cells have diameters that are large relative to ionic sizes. For this reason, the cell surface can be approximated as a planar surface.

The problem is simplified still more by assuming that the "sheets" of excess charge will be homogeneous in the two directions parallel to the membrane surface. In other words, the ions in each parallel sheet will be completely homogeneous, and the surface density ρ in each sheet will be determined only by its distance x from the membrane surface. With these assumptions, Poisson's equation is

$$d^2V/dx^2 = -\rho(x)/\varepsilon_0\varepsilon_r \qquad (8.8)$$

The density ρ is a continuous variable in this equation, so we can no longer assume discrete ions in solution. The charge is assumed to be distributed homogeneously over the surface. By the same token, the surface charges are also homogeneously distributed.

The Poisson equation provides a relationship between the potential and the excess charge at the same point. However, the excess charge is an intimate function of the potential at that point, and our analysis must include this interrelation. The excess positive charges will be attracted into the region by the negative surface-induced potential, and these charges will influence the potential. In order to relate the two parameters, the system is assumed to obey Boltzmann statistics. The energy at some distance x is determined by the potential $V(x)$ at that point, and the Boltzmann factor has the form

$$\exp[-zFV(x)/RT] \tag{8.9}$$

For cations, $z = +1$, and a negative potential at x will increase the probability of finding a cation at that position. For anions, $z = -1$, and the Boltzmann probability will decrease. The probabilities for anions and cations can be used to determine the excess of positive ionic charge at each distance x relative to the concentrations of these ions in the bulk solution where the surface potential has no effect.

The concentration of cations at a distance x from the surface is

$$c_+ = c_+^o \exp[-z_+ FV(x)/RT] \tag{8.10}$$

and the concentration of anions is

$$c_- = c_-^o \exp[-z_- FV(x)/RT] \tag{8.11}$$

where c_+^o and c_-^o are the concentrations (moles per volume) in the bulk solution. If concentrations of moles per volume are used, the Faraday constant F (coulombs/mole) must be used when moles are converted to charge for the Poisson equation. Note that the only difference between Eqs. (8.10) and (8.11) is the sign of the charge z_i; c_+ is enhanced near the membrane, while c_- is reduced.

The excess charge can be determined by taking the difference between c_+ and c_- to determine the excess moles of ion and then converting this to the excess charge. The excess density is then

$$\rho(x) = z_+ Fc_+(x) + z_- Fc_-(x) \tag{8.12}$$

A plus sign connects the two charges, because this general form simply sums the charge in the region. The negative sign of z_- insures that the negative charge is subtracted from the total positive charge at distance x.

Equations (8.10) and (8.11) can now be substituted into Eq. (8.12),

$$\rho(x) = F\{z_+ c_+^o \exp[-z_+ FV(x)/RT] + z_- c_-^o \exp[-z_- FV(x)/RT]\} \tag{8.13}$$

For a solution containing more than two ionic species, the equation can be generalized to

$$\rho(x) = F \sum z_i c_i^o \exp[-z_i F V(x)/RT] \tag{8.14}$$

When this expression for excess charge density is substituted into the Poisson equation, the resultant equation is called the Poisson–Boltzmann equation,

$$d^2 V(x)/dx^2 = -(F/\varepsilon_o \varepsilon_r) \sum z_i c_i^o \exp[-z_i F V(x)/RT] \tag{8.15}$$

For small potentials, the exponential terms in this equation can be expanded to their linear term,

$$\exp(Q) = 1 + Q \tag{8.16}$$

to produce an equation that is more easily solved. The density of excess charge is then

$$\rho(x) = F\{z_+ c_+^o [1 - z_+ F V(x)/RT] + z_- c_-^o [1 - z_- F V(x)/RT]\} \tag{8.17}$$

for the case of two ions. Collecting terms in $V(x)$ gives

$$\rho(x) = -F\{(z_+ c_+^o + z_- c_-^o) + [F V(x)/RT](z_+^2 c_+^o + z_-^2 c_-^o)\} \tag{8.18}$$

Because the term $z_+ c_+^o$ counts the total moles of positive charge in the bulk solution while $z_- c_-^o$ counts the total moles of negative charge, these two terms must sum to zero because of the electroneutrality of the bulk solution. The right-hand side of Eq. (8.15) can now be written as

$$\rho(x)/\varepsilon = -(F^2/\varepsilon RT)(z_+^2 c_+^o + z_-^2 c_-^o) V(x) \tag{8.19}$$

where

$$\varepsilon = \varepsilon_r \varepsilon_o \tag{8.20}$$

The equation can be rewritten in a more compact form by defining an ionic strength I,

$$I = (z_+^2 c_+^o + z_-^2 c_-^o)/2 \tag{8.21}$$

so that

$$\rho(x)/\varepsilon = -2(F^2 I/\varepsilon RT) V(x) = -\kappa^2 V(x) \tag{8.22}$$

All of the parameters except the position-dependent voltage have been combined into a single parameter, κ, which is called the inverse Debye length (for reasons that will become apparent next). The square of κ is chosen in Eq. (8.22) to produce a convenient form when the Poisson–Boltzmann equation is solved.

The ionic strength I weighs the ions by the square of their charge. Thus, multivalent ions will make a significantly larger contribution to the charge density than the univalent ions. The factor of 2 is introduced so that the ionic strength is equal to the concentration for univalent ions. For NaCl, with a concentration c, the ionic strength is

$$I = [(1)^2 c + (1)^2 c]/2 = c \tag{8.23}$$

For a concentration c of $LaCl_3$, the ionic strength is

$$I = [(3)^2 c + (1)^2 (3c)]/2 = 6c \tag{8.24}$$

In general, the ionic strength is

$$I = \tfrac{1}{2} \sum z_i^2 c_i^o \tag{8.25}$$

The linearized expression for the excess charge density [Eq. (8.22)] can now be substituted into the Poisson–Boltzmann equation to give

$$d^2 V/dx^2 = \kappa^2 V(x) \tag{8.26}$$

Note that the original minus in the equation and the minus generated by the exponential expansion produce a net positive on the right-hand side. The two solutions that are consistent with the form of Eq. (8.26) are

$$\exp(-\kappa x) \tag{8.27}$$

and

$$\exp(+\kappa x) \tag{8.28}$$

The general solution is a linear combination of these two,

$$V(x) = A \exp(\kappa x) + B \exp(-\kappa x) \tag{8.29}$$

where the constants A and B can be determined by introducing physically realistic boundary conditions. For example, as x approaches distances where the surface charge has no effect—that is, the bulk solution—the potential should drop to zero. Because this distance will be large compared to the molecular parameters, it is selected as $x = \infty$. The first boundary condition is then

$$V(\infty) = 0 \tag{8.30}$$

When this condition is introduced into Eq. (8.29), the term with the positive exponential becomes infinite. This condition is physically unrealistic, and the positive exponential is eliminated by setting $A = 0$. This is valid because the remaining term,

$$V(x) = B \exp(-\kappa x) \tag{8.31}$$

is still a solution of the differential equation, and the potential will become zero at large distances as required.

In order to determine a physically useful expression for B, a second boundary condition is required. If the charge per unit area on the surface is $-|\sigma|$, then the total excess charge in the solution that faces this area must be

$+|\sigma|$. We know the amount of excess charge at each value of x; that is just $\rho(x)$. By integrating over all the excess charge densities for each plane, we must recover a total charge of $+|\sigma|$. The excess ions can approach to some minimum distance a of the membrane surface and might be found at infinite distances from the surface. Thus,

$$+|\sigma| = \int_a^\infty \rho(x)\,dx \qquad (8.32)$$

This is the second boundary condition. The charge density, $\rho(x)$, has already been expressed in terms of κ,

$$\rho(x) = \varepsilon\kappa^2 V(x) = \varepsilon\kappa^2 B[\exp(-\kappa x)] \qquad (8.33)$$

so that the integral is

$$\sigma = \varepsilon\kappa^2 B \int_a^\infty \exp(-\kappa x)\,dx = \varepsilon\kappa^2 B\{\exp(-\kappa x)]/\kappa\}\,|_a^\infty$$

$$= B\varepsilon\kappa\,\exp(-\kappa a) \qquad (8.34)$$

The constant B can now be expressed in terms of the other known parameters,

$$B = \sigma\,\exp(\kappa a)/\kappa\varepsilon \qquad (8.35)$$

When this is substituted into Eq. (8.31), the equation for the double layer potential becomes

$$V(x) = (\sigma/\varepsilon\kappa)\exp[-\kappa(x - a)] \qquad (8.36)$$

The diffuse region of charge imbalance is reflected in the exponential decay of the potential as a function of distance; κ determines how rapidly the decay occurs; and κ^{-1} has the units of length and is called the Debye length. Because it represents the distance in solution where the potential has dropped to $1/e$ of its potential at the surface, it provides a measure of the effectiveness of the solution in cancelling the surface charge. Because κ is proportional to the ionic strength of the solution, κ^{-1} will be shorter when the solution concentration is larger, as there are sufficient ions available in solution to cancel the fixed surface charge in a shorter distance. Noting the definition of ionic strength, multivalent ions are more effective in cancelling this surface charge.

The Debye length, κ^{-1}, bears some similarities to the distance d that separated the two charge layers in the Helmholtz theory. The potential at $x = a$ in the Gouy–Chapman theory is

$$V(a) = (\sigma/\varepsilon\kappa) = \sigma\kappa^{-1}/\varepsilon \qquad (8.37)$$

which can be compared with the expression for the Helmholtz double layer potential,

$$V = \sigma d/\varepsilon \qquad (8.38)$$

Thus, κ^{-1} in the Gouy–Chapman theory plays the role of the fixed distance d in the Helmholtz theory. Note, however, that the use of κ^{-1} from the Gouy–Chapman theory produces the potential at the membrane surface rather than the Debye length, κ^{-1}, from the membrane surface.

3. THE GRAHAME EQUATION

The Poisson–Boltzmann equation was solved in the previous section by linearizing the Boltzmann exponentials so that the final expression is applicable when the potentials are small. In the one-dimensional case, however, the Poisson–Boltzmann equation can be solved without linearization, and the resultant exact expression provides some insights into the validity of the linearization approximation.

The Poisson–Boltzmann equation is

$$d^2V(x)/dx^2 = -(F/\varepsilon) \sum z_i c_i^o \exp[-z_i FV(x)/RT] \qquad (8.39)$$

Although this equation looks formidable, the second derivative can be rewritten as a first derivative of the potential via the identity

$$d^2V/dx^2 = (d/dV)[(dV/dx)^2]/2 \qquad (8.40)$$

This identity permits a straightforward first integration of the differential equation with respect to dV,

$$\frac{1}{2} \int_\infty^x (d/dV)[(dV/dx)^2]\, dV = -(F/\varepsilon) \int_0^{V(x)} \sum c_i^o z_i \exp(-z_i FV/RT)\, dV$$

$$-\tfrac{1}{2}(dV/dx)^2|_{x=\infty} + \tfrac{1}{2}(dV/dx)^2|_x = -(F/\varepsilon) \sum c_i^o z_i(-RT/z_i F)$$
$$\times \exp(-z_i FV/RT)|_0^x$$

$$(8.42)$$

At large distances from the surface, the potential must approach zero. This indicates that the slope dV/dx must also approach zero at these large distances, and the first derivative on the left-hand side of Eq. (8.42) must be zero. The equation is then

$$\tfrac{1}{2}(dV/dx)^2 = (RT/\varepsilon) \sum c_i^o\{\exp[-z_i FV(x)/RT] - 1\} \qquad (8.43)$$

Note the cancellation of z_i on the right-hand side. By taking the square root of both sides, the equation becomes

$$dV(x)/dx = [(2RT/\varepsilon)(\sum c_i^o\{\exp[-z_i FV(x)/RT] - 1\})]^{1/2}. \qquad (8.44)$$

A second integration of Eq. (8.44) to determine the potential would be difficult. However, this integration can be avoided by utilizing the fact that the derivative of the potential at x is the electric field at this point. This electric field can also be determined using Gauss's law, which permits us to express the electric field in terms of the charge in the region of interest. Consider a column of solution extending outward from a unit area of membrane, as shown in Fig. 8.5. If this column extended to infinity, it would contain the total excess charge, σ. Consider, instead, a new charge quantity, $\eta(x)$, which is the sum of all excess charge in the column from infinity to some distance x from the surface. This, of course, is the same regime that was used for the integration of the Poisson–Boltzmann equation.

Gauss's law defines the electric field at distance x as

$$\int \mathbf{E} \cdot d\mathbf{S} = Q_t/\varepsilon \tag{8.45}$$

The surface is defined by the cylinder of excess charge extending from x to infinity. Because of the symmetry of the cylinder, no net electric field will emanate from the sides of the cylinder: that is, electric field along $+z$ would be cancelled by the electric field along $-z$. At $x = \infty$, no electric field is expected because the solution is homogeneous. However, there will be a nonzero electric field emanating from the face of the cylinder at x, due

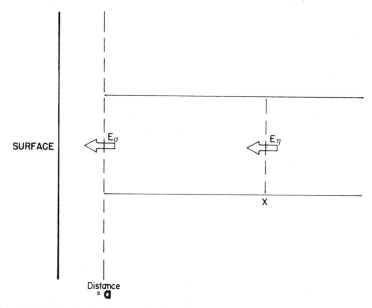

Fig. 8.5 The electric field produced by net charge within cylinders of unit area at distances $x(E_\eta)$ and $a(E_\sigma)$ from the membrane surface.

to the excess charge within the cylinder. This surface has unit area, and the total charge within the cylinder is $\eta(x)$. Thus, Gauss's law reduces to

$$-dV/dx = E = \eta(x)/\varepsilon \qquad (8.46)$$

With this relationship, integration is now unnecessary. The electric field is expressed in terms of charge in the cylinder,

$$(\eta/\varepsilon) = [(2RT/\varepsilon) \sum c_i^o \{\exp[-z_i F v(x)/RT] - 1\}]^{1/2} \qquad (8.47)$$

$$\eta = [(2RT\varepsilon) \sum c_i^o \{\exp[-z_i F V(x)/RT] - 1\}]^{1/2} \qquad (8.48)$$

This equation still contains two unknowns, η and $V(x)$.

The cylinder in Eq. (8.48) can be extended all the way to the distance of closest approach, a. The charge $\eta(x)$ will then become σ, the total excess charge in solution, and the potential $V(a)$ will be the potential at the distance of closest approach, such that

$$\sigma = [(2RT\varepsilon) \sum c_i^o \{\exp[-z_i F V(a)/RT] - 1\}]^{1/2} \qquad (8.49)$$

The equation can be simplified if the salt is a $z:z$ electrolyte, such as KCl or $CaSO_4$. The charge density is rewritten as

$$\eta = A(\sum c_i^o \{\exp[-z_i F V(x)/RT] - 1\})^{1/2} \qquad (8.50)$$

where

$$A = (2RT/\varepsilon)^{1/2} \qquad (8.51)$$

For the single $z:z$ electrolyte, all terms in the sum can be written in full to give

$$A(c^o)^{1/2} \{\exp[+|z|F V(x)/RT] - 1 + \exp[-|z|F V(x)/RT] - 1\}^{1/2} \quad (8.52)$$

which is equivalent to

$$A(c^o)^{1/2} (\{\exp[+|z|F V(x)/2RT] - \exp[-|z|F V(x)/2RT]\}^2)^{1/2}$$
$$= A(c^o)^{1/2} \{\exp[+|z|F V(x)/2RT] - \exp[-|z|F V(x)/2RT]\} \qquad (8.53)$$

$$= 2A(c^o)^{1/2} \sinh[|z|F V(x)/2RT] \qquad (8.54)$$

because

$$\sinh(x) = [\exp(x) - \exp(-x)]/2 \qquad (8.55)$$

The constants in A can be combined to give an expression for the excess charge in $\mu C/cm^2$, that is,

$$\eta(x) = 11.72(c^o)^{1/2} \sinh(19.46zV) \quad \mu C/cm^2 \qquad (8.56)$$

at 25°C. The potential at the distance of closest approach is

$$\sigma = 11.72(c^o)^{1/2} \sinh[19.46zV(a)] \quad \mu C/cm^2 \qquad (8.57)$$

where c^o has molecular units.

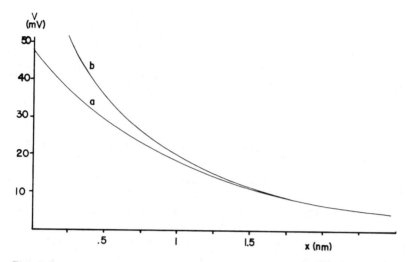

Fig. 8.6   Double-layer potential versus distance from a membrane surface for a 1:1 electrolyte in solution: line a, the Gouy–Chapman model; line b, the Grahame model.

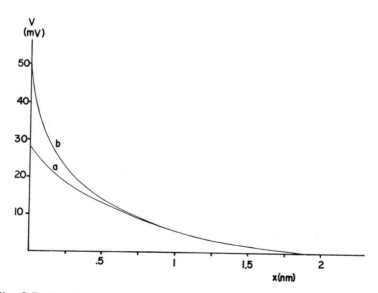

Fig. 8.7   Double-layer potential versus distance from a membrane surface for a 3:3 electrolyte in solution: line a, the Gouy–Chapman model; line b, the Grahame model.

Potentials as a function of distance x from the membrane for both the Gouy–Chapman and the Grahame equations are compared for a $1 M$ $1:1$ electrolyte, to illustrate the deviations produced by the linearization approximation, in Fig. 8.6. A $3:3$ electrolyte, $1 M$ solution is plotted for both models in Fig. 8.7, to illustrate the significant difference when $z = 3$.

4. MODIFICATIONS OF THE DOUBLE-LAYER MODELS

In our derivations of both the Gouy–Chapman and Grahame equations, a distance of closest approach a was defined. This distance might represent the approach of an ion that retained its hydration sheath or the approach of a bare ion. Grahame chose these two possibilities and assumed that each might be possible at the surface. If the ion lost its waters of hydration and approaches to roughly its ionic distance from the membrane surface, it entered a surface region that was called the inner Helmholtz plane (IHP). If the ion retained its waters of hydration, its distance of closest approach was defined by a second plane further from the surface. This second plane was called the outer Helmholtz plane (OHP). For the OHP, the ions are still essentially in an aqueous environment because of their hydration sheath. For the IHP, the ions might be considered adsorbed. This is called contact or specific adsorption and will occur when the ion is able to discard its hydration sheath. Note, however, that the surface may also have adsorbed water molecules, which will be adjacent to these adsorbed ions. This situation is considered distinct from the hydrated ionic species.

The OHP merely represents the closest that the hydrated ions can approach to the surface. Because of thermal motions, these ions will be distributed in a diffuse double layer configuration. A similar situation is impossible for ions in the IHP. If they are desorbed, they move into the OHP region and become part of the diffuse double layer. Thus, the IHP is defined much more stringently and can be accurately approximated as a Helmholtz type of double layer. Thus, the actual double layer might well be some combination of both the Helmholtz and Gouy–Chapman double-layer models.

Stern combined the two models. The IHP defined the plane for the Helmholtz layer, while the OHP defined the distance of closest approach a for the diffuse double layer. In such a combined model, there are two distinct types of potential drop. Between the surface and the IHP, the potential will change linearly; from the OHP to the bulk solution, the potential will change exponentially. The potential variation is illustrated in Fig. 8.8.

If the total surface charge per unit area is σ, then the total excess charge in the double layer must also be σ. However, this charge must now be distribu-

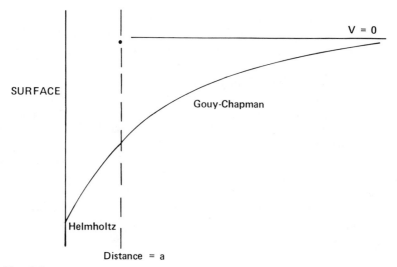

Fig. 8.8 The potential for a Stern double layer. The potential rises linearly to the fixed charge layer a and then rises toward zero exponentially in the diffuse double-layer region.

ted between the two regions, and the distribution of ions between these regions will determine the character of the potential drop. For example, if all the ions were specifically adsorbed, the model would reduce to a simple Helmholtz double layer and the potential would drop over a very short distance (approximately the combined radii of the ion and the charge containing species in the surface). A number of factors affect the magnitude of this potential. Because the distance is small, a small potential drop is predicted. However, the dielectric constant in this region is also significantly reduced, because the waters that produce the large dielectric constant in the OHP region may differ from those in the bulk solution due to the perturbing action of the surface.

Because of such complications, the exact details of capacitance and potential for the IHP and OHP will be ignored. We will assume some (inner) potential drop V_i occurs between the surface and the IHP, while an additional (outer) potential drop V_o occurs between the IHP and the OHP region. The total potential drop is then

$$V_t = V_i + V_o \qquad (8.58)$$

The potentials are those required to move a test charge from one capacitor plate to another. If the test charge is q, then the equation can be rewritten as

$$q/C_t = q/C_i + q/C_o \qquad (8.59)$$

or

$$1/C_t = 1/C_i + 1/C_o \qquad (8.60)$$

In other words, the capacitances for the two distinct double-layer regions add as if they are arranged in series. This makes intuitive sense, because the total capacitance spans a greater spatial region and would be expected to be smaller than the capacitances of its individual constituent regions.

One additional possibility must be considered for Stern models. Attraction to the surface need not be exclusively coulombic. For example, on a unit surface with 10 surface charges, it may be possible to attract 15 positively charged ions that have a strong affinity for the surface. In this case, the entire negative surface charge is neutralized and a net positive charge of 5 remains. In this case, the diffuse double layer will attract anions and exclude cations. The resultant potential variation for the double layer is illustrated in Fig. 8.9 for this case. The potential is both positive and negative over the double-layer region, and it obviously becomes more difficult to define capacitance in the conventional electrical sense, although charge is certainly being "stored" by the system.

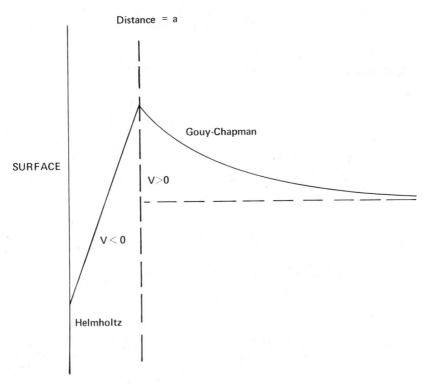

Fig. 8.9   The potential for a Stern double layer when an excess of positive charge enters the fixed charge region.

In order to develop a more general relationship that includes the effects of adsorption, we must consider how the density of surface charge affects the amount of material that is adsorbed. To do this, we must consider two types of capacitance. In calculations thus far, an integral capacitance has been used. The charge is divided by the net potential difference across the layer, that is,

$$C = q/\Delta V. \tag{8.61}$$

The alternate type, the differential capacitance, is defined as

$$C' = (\partial q/\partial V) \tag{8.62}$$

The two definitions can be compared by comparing the capacitances computed for the Grahame equation,

$$\eta(x) = A(c^o)^{1/2} \sinh[zFV(x)/2RT] \tag{8.63}$$

The differential capacitance is

$$\partial\eta(x)/\partial V(x) = A(c^o)^{1/2}(zF/2RT)\cosh[zFV(x)/2RT] \tag{8.64}$$

while the integral capacitance is

$$\eta(x)/V(x) = [A(c^o)^{1/2}/V(x)]\sinh[zFV(x)/2RT] \tag{8.65}$$

The capacitances have a completely different functional form in this case.

Consider a total potential difference V_t across the double layer. The drop will take place across the two regions, so we define the drop in terms of the two integral capacitances,

$$V_t = \sigma_\tau/C_i + \sigma_d/C_o \tag{8.66}$$

where σ_t is the total surface charge and σ_d is the surface charge that would be found for the diffuse double layer. The differential capacitance for the total double layer is now defined as an inverse, that is,

$$1/C'_t = \partial V_t/\partial\sigma_t = 1/C_t + (1/C_o)(\partial\sigma_d/\partial\sigma_t) \tag{8.67}$$

Because

$$\sigma_t = \sigma_d + \sigma_h \tag{8.68}$$

where σ_h is the cationic charge in the Helmholtz layer,

$$\partial\sigma_d/\partial\sigma_t = 1 - \partial\sigma_h/\partial\sigma_t \tag{8.69a}$$

and

$$1/C'_t = (1/C_i + 1/C_o) - (1/C_o)(\partial\sigma_h/\partial\sigma_t) \tag{8.69b}$$

The first set of bracketed terms is just the total integral capacitance resulting from the parallel addition of the capacitances. The second term suggests that this capacitance can be altered if the amount of material adsorbed changes as the surface charge changes, reflecting the fact that the total capacitance will change as the amount of adsorbed material changes.

Equation (8.69b) can be examined in some limiting cases. For example, if all the excess charge is concentrated in the Helmholtz layer, addition of 1 surface charge will add 1 charge to the Helmholtz layer. The derivative will be 1, and the total capacitance will simply remain the capacitance of the IHP. For the case of zero adsorption, both the derivative and the C_i term will be absent and the capacitance will reduce to that of the OHP.

5. SCREENING

The problem of utilizing the effects of both the IHP and OHP is obviously a complicated one, and it becomes useful to explore alternative approaches to the binding of ions to the membrane surface. One qualitative point was apparent from the previous discussion. Because the double-layer capacitances add in parallel, the larger capacitances will play the smaller role in determination of the total double layer capacitance. Since the IHP Helmholtz layer lies closer to the membrane, it is expected to have the larger capacitance and can be considered a correction to the diffuse double-layer capacitance. For this reason, let us consider alternate approaches that might permit us to include the effects of this inner layer of ions.

A reasonable alternative would be to exclude the double-layer IHP capacitative effects altogether and deal exclusively with the OHP. However, we must note that the total surface "seen" by the OHP when binding takes place will differ. Thus, we must calculate the amount of exposed membrane surface charge and use this in our calculations for the OHP.

A univalent cation C^+ is able to bind to the membrane surface sites. Assume an equilibrium between the bound and unbound C^+ of the form

$$C^+ + \sigma \longleftrightarrow C\sigma \tag{8.70}$$

with equilibrium constant

$$K = [C\sigma]/[C^+][\sigma] \tag{8.71}$$

The surface sites are then apportioned between bound and unbound sites,

$$\sigma_t = [\sigma] + [C\sigma] \tag{8.72}$$

The sites that bind C^+ at each instant are assumed to be neutralized and do not contribute to the double-layer potential. The surface charge density of interest is now σ, not σ_t.

The equilibrium expression is rewritten as

$$[C\sigma]/[\sigma] = K[C^+] \tag{8.73}$$

and, using Eq. (8.72), this becomes

$$(\sigma_t - [\sigma])/[\sigma] = K[C^+] \tag{8.74}$$

Solving this equation for σ yields

$$\sigma = \sigma_t/(1 + K[C^+]) \tag{8.75}$$

The number of open or "charged" binding sites in the unit membrane area will decrease as the binding of C^+ (or any other ion) increases. The increase can be induced by selecting an ion with a larger binding constant or increasing the concentration of the ion in solution.

This equation must be modified still further, because the binding constant K can also be a function of the double-layer potential at the surface. The equilibrium constant can be related to the free energy for binding via the relation

$$\Delta G^\circ = -RT \ln K \tag{8.76}$$

However, this free energy must now include the free energy generated by the electrical potential at the surface; that is, the electrochemical rather than the chemical free energy is used,

$$\Delta \tilde{G} = \Delta G^\circ + zFV(0) \tag{8.77}$$

The equilibrium constant is

$$K' = \exp(-\Delta \tilde{G}/RT) = K\{\exp[-zFV(0)/RT]\} \tag{8.78}$$

The concentration of open surface binding sites is

$$\sigma = \sigma_t/(1 + K'[C^+]) \tag{8.79}$$

or

$$\sigma = \sigma_t/[1 + (K[C^+]\{\exp[-zFV(0)/RT]\})] \tag{8.80}$$

In all the variations of double-layer theory discussed thus far, one central problem consistently appears. The equation involves two variables that must be determined from experiment, the surface potential V and the surface charge density σ. For this reason, it is sometimes more convenient to obtain relative values rather than absolute values for these parameters. For example, say we wished to study the effects of ions of different valency on the surface of some membrane. The total surface charge density should be the same if binding is minimal. We could then determine which concentrations of each ion produce the same potential *change*. We need not measure this potential change directly. If the potential change affects some property of the membrane, such as some kinetic process, the concentrations could be adjusted to produce the same changes in the property and this would serve as a test of the model.

For this study, an alternative form of the Grahame equation is used: this form gives the surface charge density in number of charges (not coulombs of charge) per square angstrom. The equation is now

$$\sigma = (1/272)(\sum c_i^\circ \{\exp[-z_i FV(0)/RT] - 1\})^{1/2} \quad \text{charges/Å}^2 \quad (8.81)$$

For a large potential the exponential will dominate the terms, and the expression for a univalent cation is

$$\sigma = (1/272)(c_1 \{\exp[-FV(0)/RT]\})^{1/2} \quad\quad\quad (8.82)$$

Note that the anionic term is also deleted because the exponential argument will be negative and this exponential will be small. Equation (8.82) can be rearranged as

$$\exp[-FV(0)/RT] = (272)^2/c_1 \quad\quad\quad (8.83)$$

For a divalent cation and the same approximations, the corresponding equation is

$$\exp[-FV(0)/RT] = (272\sigma)/(c_2)^{1/2} \quad\quad\quad (8.84)$$

where c_2 is the divalent cation concentration.

The left-hand sides of these two equations are identical when the ions produce identical surface potentials. For this condition, the two equations can be equated to provide a measure of the relative amounts of each cation that are required to produce this potential change,

$$(272\sigma)^2/c_1 = (272\sigma)/(c_2)^{1/2} \qu\quad\quad (8.85)$$

and

$$c_2 = (c_1)^2/(272\sigma)^2 \qu\quad\quad\quad (8.86)$$

The relative concentrations depend on the density of surface charges. As the membrane becomes significantly more charged, very little divalent ion is required to produce a voltage equivalent to that produced by univalent ions. For a surface charge density of 1 negative charge per 38 Å^2, the potential produced by a 0.1 M solution of univalent ion can be duplicated with only 2×10^{-4} M of divalent cation. Remembering that the linearized Gouy–Chapman theory would predict a factor of 4, with its square root taken when comparing univalent and divalent ions, the difference in concentrations using the Grahame equation is particularly significant.

6. ELECTROCAPILLARITY

The double-layer models that have been developed usually produced two unknowns, the surface charge density and the surface potential, that had to be determined. Both parameters cannot be controlled directly via external means. Therefore, it would be advantageous to have some system that might be controlled by an external parameter, such as an applied potential. In Chapter 2, the Maxwell relations were used to develop the Lippman equation,

$$(\partial\gamma/\partial\psi)_T = -\sigma \tag{8.87}$$

where γ is the surface tension and ψ is the measurable transmembrane potential. If the potential across an interface is changed and this produces a change in the observable surface tension, then the surface charge per unit area at the interface can be determined directly. In other words, the microscopic parameter σ has now been related to two experimentally observable parameters.

The effects of applied potential on surface tension are most readily visualized in a phenomenon called electrocapillarity. The experimental arrangement is shown in Fig. 8.10. An ionic solution lies between a pool of mercury (or a reference electrode) and a fine capillary of mercury, as shown. The mercury is an almost ideal polarizable electrode. As noted in Chapter 5, such an electrode will assume the potential applied from the source, and the applied potential is an accurate measure of the potential across the interface. If the applied voltage is zero, the negative surface tension of the mercury will cause it to move away from the tip of the tube. The location of this meniscus can be observed and is a direct measure of the surface tension of the interface.

The pressure difference across the interface separating the mercury from air is just

$$\Delta P = \rho g h \tag{8.88}$$

where the opposing density of air has been ignored relative to the high density of mercury, and h is the height of the column.

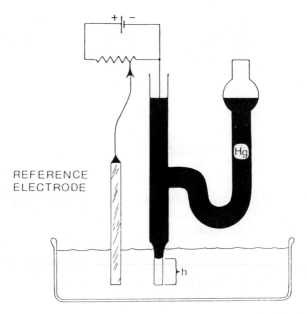

Fig. 8.10 Experimental arrangement to measure electrocapillarity. The height h in the capillary is measured as a function of the applied potential.

This pressure is opposed by the pressure produced by the curved interface,

$$\Delta P = 2\gamma/r \tag{8.89}$$

Equating these opposing pressures yields

$$\gamma = \rho ghr/2 \tag{8.90}$$

Because the radius r and ρ and g are constants, h is the only variable in this expression. Mercury produces a capillary depression rather than a capillary rise, so the height h will describe the rise of air into the capillary or the motion of the mercury into the tube.

The mercury in relatively nonreactive and polarizable. When a potential is applied, it will simply add positive or negative charge to the electrode. The surface charge density can be determined using the Lippman equation. The surface tension is recorded for each applied potential to produce a curve such as that of Fig. 8.11. By taking the slope of this curve at any potential, the Lippman equation can be used to determine the amount of excess surface charge at that potential. The curve of Fig. 8.11 passes through a maximum, so that the excess charge changes from positive to negative as a function of potential. At the maximum, the applied potential has provided just enough

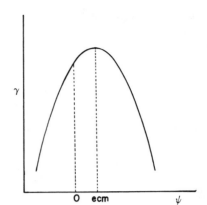

Fig. 8.11 A plot of surface tension γ versus applied potential. The peak of the parabolic curve defines the electrocapillarity maximum (ecm).

charge to balance any charge which was present. This is called the potential of zero charge.

Because both the charge and potential are now determined for each point on the curve, the differential capacity of the interface can also be defined for this system,

$$C' = (\partial\sigma/\partial\psi) = -(\partial^2\gamma/\partial\psi^2) \tag{8.91}$$

Note that the curve of Fig. 8.11 is an almost perfect inverted parabola. Because the second derivative of a parabola is a constant, a perfect parabolic shape would indicate that the capacity of interface was a constant. The system could then be modelled with the simple Helmholtz double-layer model. The first and second derivative (σ and C') curves for the curve of Fig. 8.11 are shown in Fig. 8.12.

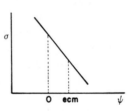

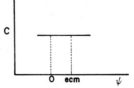

Fig. 8.12 (a) Surface charge curve and (b) capacitance curve versus applied potential, obtained as the first and second derivatives of the curve of Fig. 8.11.

The parabolic nature of the curve of surface tension versus potential can be illustrated using the Helmholtz double-layer potential expression,

$$\sigma = \varepsilon V/d \tag{8.92}$$

and the Lippman equation

$$(\partial\gamma/\partial V) = -\sigma \tag{8.93}$$

because

$$d\gamma = -\sigma\,dV \tag{8.94}$$

Substituting for σ using Eq. (8.92) gives

$$\int d\gamma = -(\varepsilon/d)\int V\,dV \tag{8.95}$$

and

$$\gamma = -(\varepsilon/d)V^2/2 + \text{constant} \tag{8.96}$$

If the surface tension is γ_0 when $V = 0$, then

$$\gamma = (\gamma_0) - \tfrac{1}{2}(\varepsilon/d)V^2 \tag{8.97}$$

which is the equation for an inverted parabola.

7. ELECTROPHORESIS AND ELECTROOSMOSIS

Although we have derived a variety of equations for the double layer, the equations and their associated experiments have been essentially static. Ions moved in and out of the double-layer region because of thermal agitation, but the relative concentrations of excess ions in the planes parallel to the surface remained constant. In addition, the lateral motion of ions within a plane did not enter the calculations.

In a series of phenomena known collectively as electrokinetic phenomena, we observe lateral motions of the ions and the charged surfaces. For example, consider a long capillary with charged walls, as illustrated in Fig. 8.13. If a potential difference is applied to the solutions on either side of the capillary, an ohmic ionic flow is expected. However, the applied potential also produces a flow of water through the capillary. This is the phenomenon of electroosmosis. The water flow can be related to the surface charge and the double layer potential generated by this charge.

To begin the analysis, we consider the complement of electroosmosis. If a charged, *mobile* particle is added to solution and a potential is applied

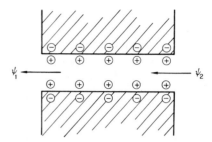

Fig. 8.13 A capillary with negative surface charge for electroosmosis.

across this solution, the particle will begin to move relative to its double layer of excess ions in solution. Note the similarity. In electroosmosis, the solution moves past a surface with fixed charge. In electrophoresis, the surface with fixed charge moves through the solution.

The velocity of the large, charged particle in solution is limited by the friction it encounters. This friction develops because the layers of water surrounding the particle move with different velocities. The layer closest to the molecule will move at the same speed as the molecule. The next layer will move more slowly, and so on. At some distance from the surface, the water layers will be unaffected by the passage of the charged particle, as shown in Fig. 8.14. What is the thickness of this frictional layer? Consider the double layer which surrounds the molecule. The solution portion of the double layer will have excess charge that is pulled in a direction opposite to that of the molecule. The double-layer region is then the area where friction will develop when the potential is applied.

Friction is observed when two solution layers move at different velocities; as we move from the membrane surface (direction x), the velocity must change. This is then represented as a derivative,

$$dv/dx \tag{8.98}$$

and the frictional force for a moving plane of unit area is

$$F = -\eta \, dv/dx \tag{8.99}$$

where η is the frictional coefficient.

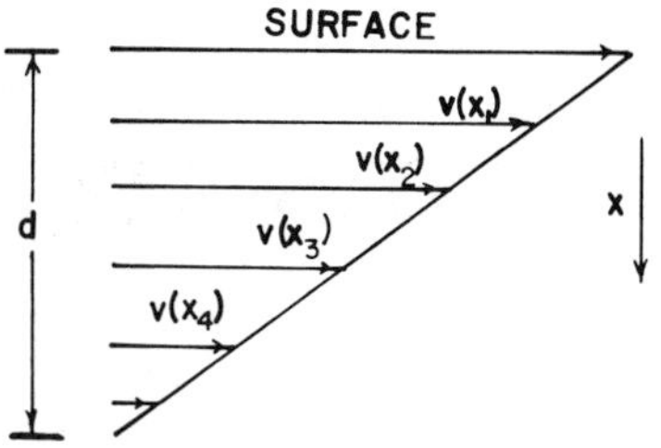

Fig. 8.14 Shear velocities of double-layer surface at increasing distances from a charged molecule moving with velocity v.

The analysis of this problem can be simplified if we assume that the planes of solution have velocities that vary linearly across the double-layer thickness d. The particle itself will be moving with some velocity v in the applied electric field. At the distance d from the surface, the solution will not be moving at all. Because of the linear variation, the derivative dv/dx is replaced by the change in velocity divided by the thickness,

$$dv/dx = \Delta v/\Delta x = (v - 0)/d = v/d \qquad (8.100)$$

and the frictional force exerted by the double-layer region is

$$F = \eta(v/d) \qquad (8.101)$$

The frictional force must oppose the motion of the charged molecule caused by the electric field. In the same unit area of molecule, there will be σ surface charges that must move relative to the solution. The total driving force is then

$$F_{dr} = E\sigma \qquad (8.102)$$

The two forces must balance when the particle moves at a constant velocity

$$\eta(v/d) = E\sigma \qquad (8.103)$$

and the mobility, v/E, is

$$u' = v/E = d\sigma/\eta \qquad (8.104)$$

where the u' represents a velocity per unit field rather than unit force. This distinction is discussed at length in Chapter 9.

The observable mobility of the particle can now be related to the double-layer potential using the Helmholtz double-layer potential expression,

$$\sigma = \varepsilon V/d = \varepsilon\zeta/d \qquad (8.105)$$

and the mobility is

$$u' = \varepsilon\zeta/\eta \qquad (8.106)$$

The zeta potential, ζ, is often used to define the double-layer potential determined from electrokinetic measurements; ε is the product of the electric permittivity and the relative dielectric constant. In cgs units, the expression is

$$u' = \varepsilon_r\zeta/4\pi\eta \qquad (8.107)$$

which is generated by dividing Eq. (8.106) by $4\pi\varepsilon_0$.

The mobility observed for a zeta potential of 100 mV in water ($\varepsilon_r = 80$) is

$$\begin{aligned} u' &= \varepsilon_0\varepsilon_r\zeta/\eta = (8.85 \times 10^{-12})(80)(0.1)/10^{-3} \\ &= 7 \times 10^{-8}(m/s)/(V/m) \end{aligned} \qquad (8.108)$$

Note that 0.01 poise must be changed to 10^{-3} N·s/m² for this equation in MKS units. The mobility for a given zeta potential can also be calculated in water using

$$u' = \zeta/1420 \tag{8.109}$$

where ζ has units of mV and u' has units of $(\mu m/s)/(V/m)$. The temperature is 20°C; the factor is lowered to 1290 at 25°C.

Equation (8.106) is used when the size of the particle is large compared to the size of the double-layer region. When the particle has dimensions comparable to those of the double layer, two theoretical expressions are possible. For small spherical particles, Eq. (8.106) remains valid. For small cylindrical particles, however, the mobility u' is two-thirds this value,

$$u' = (2/3)\varepsilon\zeta/\eta \tag{8.110}$$

We can now return to electroosmosis (Fig. 8.13). In this case, the solution moves by the capillary walls with fixed charge. However, if the walls were moving, this would be an electrophoresis problem. Because the choice of moving phase is completely arbitrary, the mobility of the solution past the capillary walls when a field is applied is just

$$u' = \varepsilon\zeta/\eta \tag{8.111}$$

8. STREAMING POTENTIAL

In electroosmosis, a solution flow was generated when a potential was applied to solutions on opposite sides of the capillary. Alternatively, a pressure might be applied across the capillary to produce a flow of solution. In this case, a potential might be generated across the length of the capillary. Such a phenomenon is observed and the resultant potential is called the streaming potential.

If fluid is forced through a capillary, the liquid nearest the capillary walls will encounter the maximal friction, while the fluid in the center of the tube will encounter the least. Thus, there will be different velocities at each radius of fluid within the capillary. For a tube of radius a and area πa^2, the total force produced by the liquid is

$$F = \pi a^2 P \tag{8.112}$$

where P is the applied pressure difference. Consider an intermediate radius x. The change in force as we increase this radius by an amount dx will be

$$dF = d(\pi x^2 P) = 2P\pi x\, dx \tag{8.113}$$

For the column of fluid with radius x, its outer surface will encounter friction with the next ring of fluid. This frictional force is proportional to the area of this cylinder of fluid of length l, that is,

$$F_{fr} = -\eta(\text{cyl area}) \, dv/dx = -\eta(2\pi xl) \, dv/dx \qquad (8.114)$$

The change in the frictional force for a change in x is

$$dF_{fr} = -2\pi\eta l \, d(x \, dv/dx) \qquad (8.115)$$

The two differential forces are now equated for radius x and common terms are cancelled, to give

$$-\eta l \, d(x \, dv/dx) = Px \, dx \qquad (8.116)$$

This equation is integrated to give

$$-nl(x \, dv/dx) = (Px^2/2) + C_1 \qquad (8.117)$$

Dividing by x and integrating again gives

$$-nlv = (Px^2/4) + C_1(\ln x) + C_2 \qquad (8.118)$$

The constants C_1 and C_2 are evaluated by noting that (1) the velocity is finite at $x = 0$, and (2) the velocity is zero at the wall. Using (1), we note that $C_1 \ln(0)$ is infinite, not finite, so $C_1 = 0$. Using (2),

$$-nlv = Pa^2/4 + C_2 = 0 \qquad (8.119)$$

$$C_2 = -Pa^2/4 \qquad (8.120)$$

The velocity through the capillary at radius x is

$$\eta lv = (P/4)(a^2 - x^2) \qquad (8.121)$$

$$v = (P/4\eta l)(a^2 - x^2) \qquad (8.122)$$

If the double layer is located close to the surface of the capillary,

$$x = a - d \qquad (8.123)$$

Substituting into Eq. (8.122) and ignoring the term in d^2 gives

$$v = (P/4\eta l)(2ad) = Pad/2\eta l \qquad (8.124)$$

Because we now know the velocity of the solution half of the double layer, we can calculate the amount of surface charge that moves out each second. The cylindrical region that reaches solution each second has an area (Fig. 8.15) of

$$(2\pi a)v \qquad (8.125)$$

and, because σ is the charge per unit area, the total current that leaves in this time is

$$i_{st} = \sigma(2\pi a)v \qquad (8.126)$$

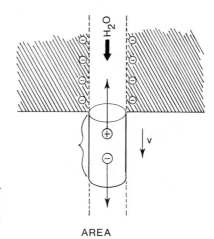

Fig. 8.15 A charged cylindrical plug of solution passing through a charged capillary. Concentric cylinders nearer the capillary fixed charges move more slowly.

Substituting Eq. (8.124) into this current expression gives

$$i_{st} = \sigma(2\pi a)(Pad/2\eta l) = \pi a^2 \sigma dP/\eta l \tag{8.127}$$

If κ is the conductivity of the solution, then the actual conductance of the capillary is

$$G = (\pi a^2)\kappa/l \tag{8.128}$$

and the streaming potential is

$$\psi = i_{st}/G = [\pi a^2 \sigma d(P)/\eta l]/[(\pi a^2 \kappa)/l] \tag{8.129}$$

$$\psi = \sigma dP/\eta \kappa \tag{8.130}$$

The Helmholtz double-layer expression is substituted to give an expression in terms of the zeta potential,

$$\psi_{st} = \varepsilon \zeta P/\eta \kappa \tag{8.131}$$

For the cgs form of this equation, divide by $4\pi\varepsilon_0$:

$$\psi_{st} = \varepsilon_r \zeta P/4\pi\eta \kappa \tag{8.132}$$

Note that both the streaming current and streaming potential are directly proportional to the pressure P. Such linear relationships will be generalized for the study of irreversible thermodynamics in Chapter 10.

To derive the expression for streaming potential and streaming current, we forced the solution past fixed surface charges. By analogy with the mutual complementarity of electrophoresis and electroosmosis, we expect a complementary effect if charged particles are moved through the solution by some force. Such a phenomenon is observed when large charged particles sink in solution under the influence of gravity; the resultant potential is called the sedimentation potential.

The equation to describe the sedimentation potential is almost identical to the streaming potential equation. However, we must now use gravity to move the particles through the solution; the pressure in the streaming potential equation is replaced with a gravitational term. The gravitational force for a concentration of n_0 spheres per unit volume is

$$F = n_o mg \qquad (8.133)$$

where the mass m of the particle with radius a in a solution of density ρ_0 is

$$m = (4\pi a^3/3)(\rho - \rho_o) \qquad (8.134)$$

The driving "pressure" for the sedimentation potential is

$$F = n_o g(4\pi a^3/3)(\rho - \rho_o) \qquad (8.135)$$

The sedimentation potential is then

$$\psi_{\text{sed}} = \varepsilon \zeta \eta_o g(4\pi a^3/3)(\rho - \rho_o)/\eta \kappa \qquad (8.136)$$

PROBLEMS

1. Use the equation for parallel-plate capacitance [Eq. (8.3)] to verify that ε_0 has the units of $C^2/N \cdot m^2$.

2. Fixed surface charges on a membrane are separated by 1 nm. Hydrated ions in solution form a second planar layer at a distance of 0.5 nm from the surface. Determine the charge density in coulombs per square meter and the Helmholtz double-layer potential for the membrane system.

3. Determine the ionic strength of a 0.5 M solution of $La_2(SO_4)_3$ and compare it with the ionic strength of a 0.5 M solution of $LaPO_4$.

4. The double-layer potential at a is -100 mV ($RT/F = 25$ mV) when 1 mM $CaCl_2$ is present in solution. Assume that Ca^{2+} dominates the terms in the Grahame equation and make an estimate of the surface charge density. Determine the surface charge density, including both Ca^{2+} and Cl^-.

5. For the same surface charge density, will Na^+ or Ca^{2+} produce a lower absolute potential at the distance of closest approach if both are present at the same concentration? Explain.

6. A membrane has a surface charge density of 1 charge/50 Å^2. If the binding constant of a divalent cation, C^{2+}, is 0.02, and the solution concentration of C^{2+} is 0.1 M, determine the effective surface charge when the potential at the surface is zero. How will this effective surface charge change when the potential at the distance of closest approach is -50 mV?

7.   Determine the concentration of a trivalent ion, c_3, that will produce the same potential shift at the surface as a 0.1 M solution of a univalent cation. Ignore the anion effects.

8.   Determine the concentration of a trivalent ion, c_3, that produces the same potential shift as a 0.01 M solution of C^{2+} ion.

9.   A double layer with surface charge density 10^{14} charges/cm^2 generates a double-layer potential of 25 mV when the solution is 1 mM NaCl. What potential would you expect for a solution of 1 mM CaSO$_4$?

10.   How will the relative length of the Gouy–Chapman layer vary for the following solutions? (a) 1 mM NaCl; (b) 1 mM CaCl$_2$; (c) 0.5 mM CaCl$_2$ and 0.5 mM NaCl.

11.   An electroosmosis experiment is run with channels containing fixed positive charge. The zeta potential is -1 mV. Determine the mobility of water flow.

12.   Convert the solvent mobility for electroosmosis into an expression for volume flow through N channels of radius a.

13.   In a solution of LaCl$_3$, La^{3+} ion dominates the double-layer potential for an anionic membrane. When the La^{3+} concentration is 0.1 mM, a double-layer potential of 75 mV is observed at 25°C. (a) Determine the surface charge density. (b) When the concentration is decreased to 0.05 mM, the maximal double-layer potential remains at 75 mV. Determine the amount of bound La^{3+} released from the surface at the lower concentration.

Chapter 9

Diffusion

1. THE NATURE OF DIFFUSION

Thermodynamics can be used to develop interrelations between phases by utilizing the notion that such phases must be in equilibrium. The molecular details of the phases are not required, and time does not enter the calculations. The equilibrium expressions hold only after the phases have reached equilibrium; the time required to attain this equilibrium is not considered. For example, if one of two equal concentrations bathing a membrane is suddenly changed to a higher concentration, ions must move toward the lower concentration. The rate at which these ions reestablish equilibrium now depends on a variety of molecular parameters, such as the speed of the ion in solution and its speed in the membrane. To study the kinetic processes that lead to equilibrium, new information on the system is required beyond that required for thermodynamic analyses. Although this may seem a disadvantage, it is not, because kinetic experiments can be used to gather information on the molecular parameters of the system. For example, the electrokinetic phenomena of Chapter 8 provided detailed information on the microscopic surface potential of the interfacial systems. In this chapter, we will consider the rate at which particles pass through the membrane due to driving forces applied to the bathing solutions. Such processes are called diffusion processes.

To study the kinetics of diffusion, this kinetics must occur on an observable time scale, which must be short relative to the time required to establish equilibrium. Such time scales can be controlled by a proper choice of membrane bathing solutions. For example, if the bathing solutions contain 0.1 mole of diffusing particles and the membrane restricts flow to only 10^{-10}

mole, a considerable time must pass before the number of particles in the bath is affected. Such a situation is called a steady state; particles flow through the membrane at some steady rate with minimal perturbation of the bath concentrations. If the time scale required to reach equilibrium is comparable to diffusion times through the membrane, then the kinetics will describe the temporal evolution of the system to equilibrium. Similar methods of analysis are used for both cases.

Consider the situation of Fig. 9.1, in which there are 20 particles in the left bath and 10 particles in the right bath. From our previous studies of equilibrium, we know that the system must ultimately evolve to a state in which there are 15 particles in each of these equivolume baths. How does this happen?

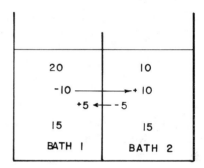

Fig. 9.1 Diffusion between Bath 1 with 20 particles and Bath 2 with 10 particles.

Assume that all the particles are in random motion. For our present purposes, this means they have an equal chance of moving either left or right. There is no preferred direction. How can such random motions be converted into a flow to the right? The source of the directional flow lies in the relative numbers of particles; on the left side at some instant, half (10) of the particles will be moving left. The remaining half are moving right and can cross the boundary. On the right side, half (5) of the particles are moving left across the boundary on the average. After these particles cross the boundary, the right side will have the final total

$$10 + 10(\text{from left}) - 5(\text{lost from right}) = 15 \qquad (9.1)$$

while the left side has the net change

$$20 - 10(\text{to right}) + 5(\text{from right}) = 15 \qquad (9.2)$$

Thus, the flow of particles through the membrane will be dictated by the number of particles on each side, and we must develop mathematical expressions that quantify this idea.

2. FORCE–FLUX RELATIONS

Although our major emphasis in this chapter centers on the diffusion of particles through membranes, it is convenient to start the discussion with a more common system, which establishes the general form of the equations which will be used. When a potential is applied to some conductor, electrons will flow in this conductor; the relationship is linear and is called Ohm's law,

$$i = GV \qquad (9.3)$$

where i is the current, V is the applied potential difference, and G is a conductance for the medium. Obviously, the amount of current will depend on properties of the conductor such as length and cross-sectional area. The expression can be written in a more general form by introducing the conductivity, which is simply the conductance through a unit cube of the material; the conductivity κ has units of $(\text{ohm m})^{-1}$. The conductance and conductivity are related through the length and area of the actual conductor under study,

$$G = \kappa A/L \qquad (9.4)$$

This expression can now be substituted into Eq. (9.3) to produce a more general form of Ohm's law,

$$i = (\kappa A/L)V \qquad (9.5)$$

or

$$J = i/A = \kappa(V/L) = \kappa E \qquad (9.6)$$

where J, the flux, is the current through a unit area of the conductor and E is the electric field across the conductor. Equation (9.6) is a vector equation; a field in a given direction produces a flux of charge in that direction.

When Ohm's law is written in this more general form, it suggests the type of relationship that might be used for other flow processes. For example, the driving force, the electric field, is the derivative of the potential applied to the system. To illustrate this distinction, consider two plates that are separated and have a potential difference V. The distance between the plates has no effect on the potential difference. However, it does determine the magnitude of the electric field. If the plates are close enough together, the resultant large electric field produces ionization of the air and a current flow, that is, flux.

Fourier's law of heat flow is a second example of a force–flux equation. The flow of heat is related to the temperature gradient,

$$J_q = -\lambda \, dT/dx \qquad (9.7)$$

In this case, the temperature represents the potential for heat flow but the gradient determines the speed of flow. If a hot object is placed next to a cold object, it will cool very rapidly. As the objects are separated, the heat flow decreases accordingly.

For a diffusion process, the potential of interest is the chemical potential,

$$d\mu = RT \, d(\ln c) \tag{9.8}$$

By analogy with previous examples, the driving force for diffusion will be the derivative of this potential in the direction of flow, that is, the x direction,

$$X_{\text{dif}} = -d\mu/dx = RT \, d(\ln c)/dx = -(RT/c)(dc/dx) \tag{9.9}$$

As expected, the driving force for diffusion depends on how rapidly the concentration changes with distance, dc/dx.

By analogy with the previous examples, a force–flux relationship is postulated for diffusion,

$$J_p = -D \, dc/dx \tag{9.10}$$

where the proportionality constant, D, is the diffusion coefficient with units of m^2/s (cm^2/s in cgs units). This choice of units provides consistent dimensions on each side of the equation. Equation (9.10) is Fick's first law of diffusion.

3. FICK'S FIRST LAW OF DIFFUSION

Fick's first law appears as a differential equation, which can now be solved for some simple cases to illustrate how it describes the motions of particles through a membrane. Consider the case of flow through a homogeneous membrane when Bath 1 has a concentration c_1 and Bath 2 has a concentration c_2, as shown in Fig. 9.2. The membrane has a thickness L, and we consider

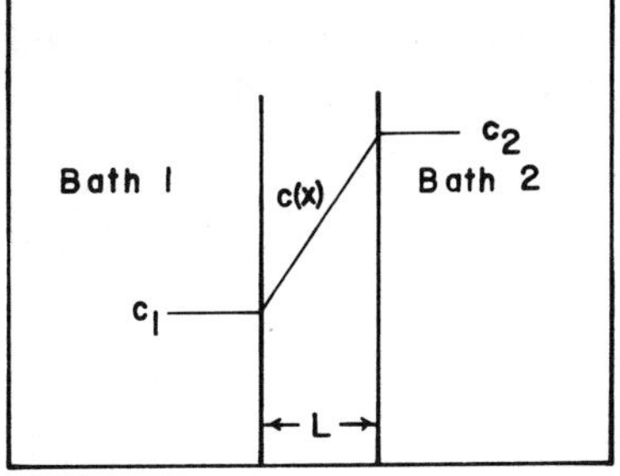

Fig. 9.2 The linear concentration gradient in a membrane separating baths of concentrations c_1 and c_2, respectively.

only the flow through some unit cross-sectional area so that the resultant flow is equivalent to the flux. Baths 1 and 2 are assumed large, so that their concentrations are essentially constant on the experimental time scale.

If the system is in a steady state, there is a constant flux J of particles through the membrane. Fick's first law is

$$J\,dx = -D\,dc \tag{9.11}$$

Since both J and D are constants, the differentials dc and dx are immediately integrated over the region of interest. At $x = 0$, $c = c_1$, and at $x = L$, $c = c_2$, so the solution is

$$Jx\Big|_0^L = -Dc\Big|_{c_1}^{c_2} \tag{9.12}$$

$$JL = -D(c_2 - c_1) \tag{9.13}$$

which is now solved for the net flux through the membrane,

$$J = -(D/L)(c_2 - c_1) \tag{9.14}$$

Thus, net flux is proportional to the concentration difference of the two bathing solutions. The proportionality constant is the diffusion coefficient divided by the membrane thickness. For many systems, it is impossible to measure the diffusion coefficient and membrane thickness independently, and it is convenient to define a permeability coefficient P such that

$$P = D/L \tag{9.15}$$

with units of m/s. Although these units are the same as those of a velocity, the permeability is not strictly a velocity, although it does provide a measure of the speed with which the particles pass through the membrane.

Although the transmembrane flux is normally the experimental parameter of interest, there is some physical interest in knowing the concentrations of particles at different points within the membrane during the steady-state diffusion process. In order to determine such concentrations, Fick's law is integrated to some arbitrary distance x within the membrane. At this point, the concentration will be c. The integration is identical although the limits are now different,

$$Jx'\Big|_0^x = -Dc'\Big|_{c_1}^{c_2} \tag{9.16}$$

and

$$Jx = -D(c - c_1) \tag{9.17}$$

Because J is constant and has been calculated previously [Eq. (9.13)], it can be substituted into Eq. (9.16) to give

$$x(-D/L)(c_2 - c_1) = -D(c - c_1) \qquad (9.18)$$

which can be rearranged and solved for c,

$$c = c_1 + (x/L)(c_2 - c_1) \qquad (9.19)$$

The concentration increases linearly from c_1 to c_2 as x increases across the membrane, as shown in Fig. 9.2. The steady flow of particles through the membrane establishes a concentration distribution, $c(x)$, in the membrane. At equilibrium, a homogeneous distribution of particles would be expected.

4. FICK'S SECOND LAW AND THE EQUATION OF CONTINUITY

The constant flux that appeared in Fick's first law requires a steady flow of particles through the membrane. However, this is restrictive. Consider the example of the previous section and assume that, initially, both sides have concentration c_1. When Bath 2 concentration is suddenly increased to c_2, the linear concentration distribution cannot be established immediately. Particles must flow into the membrane to establish the distribution, and this requires time. A more general differential equation is required to describe such processes.

To generalize Fick's first law, time must be introduced as an explicit variable and flux must also become a variable. When J was constant, we could select an arbitrary slice of membrane (Fig. 9.3). As J particles crossed the left boundary, J particles would be leaving the region via the right boundary. The particles entering and leaving could be different, but there could be no net change in the number of particles in the slice.

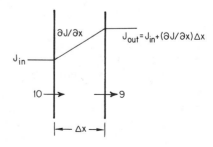

Fig. 9.3 Variable flux of particles across a region x. Particles that leave the region more slowly produce a net flux difference between the boundaries.

Consider the situation where the flux is not constant across the membrane. If 10 particles entered the slice via the left boundary and only 9 left via the right boundary in the same period of time, the missing particle will be left in the region. Thus the concentration of the region would increase with time and the rate of change would be related to the change of flux with distance.

The net accumulation of particles in a slice of thickness Δx will be the flux in, less the flux out, multiplied by the time Δt of the observation:

$$(J_x - J_{x+\Delta x})\,\Delta t \qquad (9.20)$$

For small values of x, $J_{x+\Delta x}$ can be expressed with respect to J_x by expanding it in a Taylor series about J_x,

$$J_{x+\Delta x} = J_x + (dJ_x/dx)\,\Delta x \qquad (9.21)$$

The flux difference is

$$J_x - J_x - (dJ_x/dx)\,\Delta x = -(dJ_x/dx)\,\Delta x \qquad (9.22)$$

The accumulation of particles is then directly related to the variation of the flux across the slice. Because we are considering a region of unit area, the total change in the number of particles in the slice will be

$$\Delta c(1\text{ m}^2)\,\Delta x \qquad (9.23)$$

where c is the concentration and $1(\Delta x)$ is the volume of the slice. We can now equate the two expressions for the accumulation of particles in the slice

$$\Delta c\Delta x = -[dJ(x)/dx]\,\Delta x\,\Delta t \qquad (9.24)$$

The Δx terms cancel and, in the limit of short times, dt, the equation reduces to

$$(\partial c/\partial t) = -(\partial J/\partial x) \qquad (9.25)$$

where partial derivatives have been used to emphasize the presence of two independent variables, x and t, in this equation. Equation (9.24) is an example of an equation of continuity.

The equation of continuity can be generated in three dimensions by using the divergence theorem, which was introduced in Chapter 3. For an arbitrary surface, the total flux across the boundaries is

$$-\int \mathbf{J} \cdot d\mathbf{S} \qquad (9.26)$$

which represents the sum of all fluxes across the entire enclosed surface. Since a net change inside implies the concentration is changing, the total change will be the rate of change of concentration at each point within the region integrated over the entire region,

$$\int_{V_t} (\partial c/\partial t)\,dV \qquad (9.27)$$

Because both these equations describe the same accumulation of particles, they are equated

$$\int_{V_t} (\partial c/\partial t)\, dV = -\int \mathbf{J} \cdot d\mathbf{S} \tag{9.28}$$

The divergence theorem is used to convert the surface integral to a volume integral,

$$\int_{V_t} (\partial c/\partial t)\, dV = -\int_{V_t} \mathbf{\nabla} \cdot \mathbf{J}\, dV \tag{9.29}$$

Now, however, both integrals have the same integrand and are integrated over the same region. Thus, the two functions are equal,

$$\partial c/\partial t = -\mathbf{\nabla} \cdot \mathbf{J} \tag{9.30}$$

This equation can be used with any choice of coordinate system, such as spherical coordinates.

The equation of continuity can now be used to generate Fick's second law of diffusion. Fick's first law can be written as a vector equation,

$$\mathbf{J_p} = -D\, \mathbf{\nabla} c \tag{9.31}$$

where $\mathbf{\nabla} c$ is the gradient of the concentration. We now take the divergence of both sides of Eq. (9.31),

$$\mathbf{\nabla} \cdot \mathbf{J_p} = -D\, \mathbf{\nabla} \cdot \mathbf{\nabla} c = -D\nabla^2 c \tag{9.32}$$

Using Eq. (9.30), this becomes

$$\partial c/\partial t = D\nabla^2 c \tag{9.33}$$

which is the three-dimensional form of Fick's second law. In one dimension, the del-squared term can be replaced by the second derivative of concentration with respect to x, and the equation is

$$\partial c/\partial t = D\, \partial^2 c/\partial x^2 \tag{9.34}$$

Note that the negative signs from the equation of continuity and Fick's first law cancel when the two equations are combined.

5. DIFFUSION ACROSS A MEMBRANE

In our study of Fick's first law, we examined a membrane system in which the left bath (Bath 1) had a concentration of c_1 and the right bath (Bath 2) had a concentration of c_2. The concentration varied linearly across the membrane at steady state.

Now consider the situation when both sides have concentration c_1 and, at time zero, the concentration of Bath 2 is instantaneously increased to c_2. Particles will enter the membrane from Bath 2 until the linear, steady-state condition is reestablished. We wish to follow the evolution of this process in time.

Because the baths are large, the concentration at the interfaces will not change with time. This provides some boundary conditions for the membrane system,

$$c_{(x=0)} = c_1 \qquad c_{(x=L)} = c_2 \tag{9.35}$$

At time zero, the concentration in the membrane must be c_1 because this was the concentration on both sides of the membrane.

$$c(x, 0) = c_1 \qquad \text{for all } x \tag{9.36}$$

The initial distribution in the membrane and the final steady state are illustrated in Fig. 9.4.

The boundary conditions at the membrane interfaces are c_1 and c_2. It is convenient to arrange matters such that these boundary conditions are 0. We define a new function that has zero boundary conditions and can be related to the original function $c(x, t)$. The boundaries will have the values c_1 and c_2 throughout the diffusion, so a new function is defined by subtracting the steady-state final solution from $c(x, t)$,

$$C(x, t) = c(x, t) - c_1 - (c_2 - c_1)(x/L) \tag{9.37}$$

This new function is zero at both boundaries, that is,

$$C(0, t) = C(L, t) = 0 \tag{9.38}$$

At $t = 0$,

$$C(x, 0) = c_1 - c_1 - (c_2 - c_1)(x/L) = -(c_2 - c_1)(x/L) \tag{9.39}$$

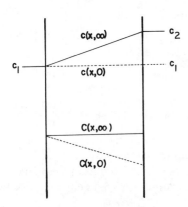

Fig. 9.4 Initial and final concentrations in a membrane in which the concentration c_2 ($c_2 > c_1$) is reduced to c_1.

Note that at long times, $c(x, t)$ approaches the steady-state solution so that $C(x, t)$ approaches zero at long times. It disappears, leaving only the steady-state solution for $c(x, t)$.

The second-order differential equation

$$\partial C(x, t)/\partial t = D\, \partial^2 C(x, t)/\partial x^2 \tag{9.40}$$

is solved using the method of separation of variables. The term $C(x, t)$ is assumed to be the product of functions that depend exclusively on x or t,

$$C(x, t) = X(x)\, T(t) \tag{9.41}$$

Substituting this function into the differential equation and dividing by the function on both sides gives

$$[1/DT(t)](dT/dt) = [1/X(x)][d^2X(x)/dx^2] \tag{9.42}$$

Since each side of this equation now includes separate, independent variables, they must both be equal to some common constant, which is chosen as $-a^2$. The two equations in x and t can now be separated,

$$dT(t)/dt = -a^2 DT(t) \tag{9.43}$$

and

$$d^2X(x)/dx^2 = -a^2 X(x) \tag{9.44}$$

Equation (9.43) will have a solution

$$T(t) = \exp(-a^2 Dt) \tag{9.45}$$

The solution to Eq. (9.44), which also contains the constants that must be determined, will be

$$X(x) = A\sin(ax) + B\cos(ax) \tag{9.46}$$

The constants A and B are determined by reference to the boundary conditions. Since $C(0, t) = 0$ and the exponential for $T(t)$ cannot be zero at all times, $X(0)$ must be zero, that is,

$$0 = A\sin(a \cdot 0) + B\cos(a \cdot 0) \tag{9.47}$$

This condition is satisfied when $B = 0$. The function is also zero at $x = L$, so that

$$0 = A\sin(aL) \tag{9.48}$$

This boundary condition can be satisfied only if

$$aL = n\pi \qquad n = 1, 2, 3, \ldots \tag{9.49}$$

and this defines a for the problem,

$$a = n\pi/L \tag{9.50}$$

The solutions for $X(x)$ and $T(t)$ are now

$$X(x) = A \sin(n\pi x/L) \tag{9.51}$$

$$T(t) = \exp[-(n\pi/L)^2 Dt] \tag{9.52}$$

The function $C(x, t)$ is

$$C(x, t) = A \sin(n\pi x/L) \exp[-(n\pi/L)^2 Dt] \tag{9.53}$$

Only the constant A must be determined by using the boundary condition for $t = 0$. Using this boundary condition,

$$C(x, 0) = A \sin(n\pi x/L) = -(c_2 - c_1)(x/L) \tag{9.54}$$

Thus, we must use sine functions with different values of n to reproduce the linear function on the right. This is an example of a Fourier series, and it is necessary to determine the coefficients A_n associated with each integer n. These Fourier coefficients have the form

$$A_n = (2/L) \int_0^L [-(c_2 - c_1)(x/L)][\sin(n\pi x/L)] \, dx \tag{9.55}$$

$$= -(2/L)(c_2 - c_1)L^{-1} \int_0^L x[\sin(n\pi x/L)] \, dx \tag{9.56}$$

$$= -(2/L)(c_2 - c_1)L^{-1}[L/n\pi]^2[\sin(y) - y\cos(y)] \Big|_0^{n\pi} \tag{9.57}$$

$$= [-(2/L)(c_2 - c_1)L/(n\pi)^2][\sin(n\pi) - n\pi \sin(n\pi) - n\pi \cos(n\pi) + 0] \tag{9.58}$$

$$= (2/n\pi)(c_2 - c_1)\cos(n\pi) \tag{9.59}$$

$$= (2/\pi)(c_2 - c_1)[(-1)^n/n] \tag{9.60}$$

The solution for $C(x, t)$ is

$$C(x, t) = T(t)X(x) = \sum \exp[-(n\pi/L)^2 Dt]A_n \sin(n\pi x/L) \tag{9.61}$$

and, since

$$c(x, t) = c_1 + (c_2 - c_1)(x/L) + C(x, t) \tag{9.62}$$

the solution for $c(x, t)$ is

$$c(x, t) = c_1 + (c_2 - c_1)(x/L) + \sum \exp[-(n\pi/L)^2 Dt]A_n \sin(n\pi x/L) \tag{9.63}$$

For long times, the exponential terms will decay to zero for each n, and $c(x, t)$ will approach its steady-state value. However, the rate at which each term in the summation decays will differ because of n^2. The argument for $n = 2$

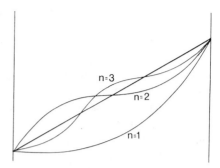

Fig. 9.5 Individual Fourier components that add to describe the concentration $c(x, t)$ within the membrane at time t.

will be four times as large as that of $n = 1$ and will disappear much more rapidly. Terms with larger n will decay even more rapidly. Over most of the diffusion process, the rate of approach to a steady state will be controlled by the $n = 1$ term. The rationale is illustrated by showing the individual terms in the series $n = 1, 2, 3$ in Fig. 9.5, and the sum of one, two, and three waves in Fig. 9.6. The higher terms in the series provide the sharp "edges"

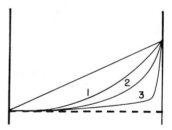

Fig. 9.6 Summation of one, two, and three Fourier components to illustrate the increased accuracy of the description of the initial state of the system.

that are expected when the diffusion first starts. After a short time, particles enter from both interfaces and smooth out the distribution. The last section to be filled is the center of the membrane, which corresponds to the maximal negative amplitude of the $n = 1$ term. The time evolution of the concentration distribution within the membrane is illustrated in Fig. 9.7.

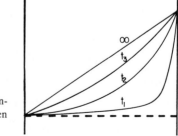

Fig. 9.7 Time evolution of the concentration $c(x, t)$ within the membrane when c_2 is set equal to c_1 at $t = 0$.

6. THE NERNST–PLANCK EQUATION

Fick's laws describe the motions of particles when a concentration gradient is present. If these diffusing particles were ions, an applied potential could also produce diffusion and the net motion would be some combination of both effects. Because we have now established the approach for creating force–flux equations, extension to this new case is straightforward. The electrochemical potential, $\hat{\mu}$, is used in place of μ so that the force–flux relation is

$$J_P \propto d\hat{\mu}/dx \qquad (9.64)$$

In previous sections, the proportionality constant was the diffusion coefficient D. In this section, we use an alternate derivation of this constant that provides some new insights into the physical nature of D.

In Eq. (9.14), the flux has the units of moles (or particles) per unit area per second. The term on the right has the units of force. To eliminate this dimension we introduce the mobility, which has the units of velocity per force. The units on each side are now

$$\text{mol/m}^2\text{s} = (?)(\text{m/s}) \qquad (9.65)$$

The dimensions become consistent only if we add an additional term with units of moles/m^3, that is, a concentration. The proportionality constant is just

$$uc \qquad (9.66)$$

and the force–flux relation is

$$J = -uc \, d\hat{\mu}/dx \qquad (9.67)$$

The factor uc has a simple physical interpretation. All the particles in some plane of unit area will move a distance equal to the particle velocity in 1 s. In other words, all the particles in a volume vA where v is the velocity will have crossed the boundary. The number of moles of particles is

$$n = cvA \qquad (9.68)$$

or

$$J = n/A = cv = cu \, d\mu/dx \qquad (9.69)$$

The force associated with the electrochemical potential is now

$$d\hat{\mu}/dx = (d/dx)[RT \, d(\ln c) + zF \, d\psi] \qquad (9.70)$$

$$= (RT/c)(dc/dx) + zF \, d\psi/dx \qquad (9.71)$$

and the force flux equation is

$$J_P = -uc[(RT/c)(dc/dx) + zF \, d\psi/dx] \tag{9.72}$$

$$= -(uRT)(dc/dx) - zFuc \, d\psi/dx \tag{9.73}$$

Equation (9.73) is the Nernst–Planck equation for electrodiffusion.

The proportionality constant that relates the flux to the concentration gradient is now independent of concentration. It has the units

$$u\left[\frac{m/s}{force}\right][force \cdot m] = \frac{m^2}{s} \tag{9.74}$$

because energy is just the product of force and distance. The dimensions of uRT are identical to those of the diffusion coefficient because they describe the same force–flux relationship. Therefore,

$$D = uRT \tag{9.75}$$

This is the Einstein equation.

Mobility must now be defined in a completely consistent way. We have used for mobility the definition velocity per unit force. The mobility is also defined as velocity per unit field, as noted in Chapter 5. In the present development, velocity per unit force has been used because we are dealing with a flow of particles. However, the particle mole flux can be converted to a charge flux by multiplication with zF:

$$I = zFJ = -zFuRT \, dc/dx \tag{9.76}$$

The charge term, zF, can be associated with the particle mobility to convert it to a charge mobility,

$$zFu = u/(1/zF) = velocity/(force/charge) = velocity/field \tag{9.77}$$

Thus, a charge flux equation is written as

$$I = -u'c \, d\mu/dx \tag{9.78}$$

where

$$u' = zFu \tag{9.79}$$

and the two definitions of mobility are consistent.

The second term in the Nernst–Planck equation is just Ohm's law. Because

$$J = \kappa E \tag{9.80}$$

the conductivity can be defined as

$$\kappa = zFuc \tag{9.81}$$

The conductivity depends on the concentration of the ion in solution; this is consistent, because the ions carry the current. The Nernst–Planck equation is simply a linear combination of Fick's and Ohm's laws.

7. THE GOLDMAN CONSTANT FIELD EQUATION

The diffusion equation for the flux J contains derivatives of both the concentration and electrical potential, and this implies that both quantities must be known as a function of position. However, a considerable simplification is possible if the functional form of one of these quantities is selected and the second quantity is deduced from it.

Goldman proposed that the electric field across the membrane be a constant—that is, that the electric field would not be affected by the presence of ions in the membrane. Under such circumstances, the electric field is defined by the applied potential and the membrane thickness and is constant. The field is

$$E = -d\psi/dx = -\psi/L \qquad (9.82)$$

where ψ is the transmembrane potential and L is the membrane thickness. This constant field is substituted into the Nernst–Planck equation to give

$$J = -D\,(dc/dx) - zFuEc = -D\,(dc/dx) - Ac \qquad (9.83)$$

where

$$A = zFuE = \text{constant} = zFu\psi/L \qquad (9.84)$$

With the constant field, the Nernst–Planck equation becomes a first-order differential equation for c,

$$-D\,(dc/dx) = J + Ac \qquad (9.85)$$

This equation must include boundary conditions. No initial time information is necessary because a steady (constant) flux is assumed. The concentrations at each boundary must be specified; we assume the concentrations just inside the membrane at each interface are equivalent to the concentrations in the adjacent bulk solution. In more involved models, a partition coefficient for the solution–membrane interface must be used. The boundary conditions are

$$x = 0 \quad \text{with} \quad \text{concentration } c^o \text{ and } \text{potential } \psi^o \qquad (9.86)$$

$$x = L \quad \text{with} \quad \text{concentration } c^L \text{ and } \text{potential } \psi^L \qquad (9.87)$$

$$\psi = \psi^L - \psi^o \quad \text{is the potential difference across the membrane} \qquad (9.88)$$

The diffusion of a single ionic species is considered.

The differential equation is rearranged as

$$\int_{c^o}^{c^L} [-D/(J + Ac)] \, dc = \int_0^L dx \tag{9.89}$$

Integration gives

$$(-D/A) \ln(J + Ac) \Big|_{c^o}^{c^L} = x \Big|_0^L \tag{9.90}$$

or

$$\ln[(J + Ac^L)/(J + Ac^o)] = -AL/D \tag{9.91}$$

Exponentiating both sides gives

$$[(J + Ac^L)/(J + Ac^o)] = \exp(-AL/D) = B \tag{9.92}$$

This equation is now solved for J as a function of the concentrations:

$$J + Ac^L = BAc^o - Ac^L$$

$$J(1 - B) = BAc^o - Ac^L$$

$$J = (BAc^o - Ac^L)/(1 - B) = A(Bc^o - c^L)/(1 - B) \tag{9.93}$$

By multiplying the numerator and denominator by B^{-1}, the equation becomes

$$J = -A(c^o - c^L B^{-1})/(1 - B^{-1}) \tag{9.94}$$

and,

$$B^{-1} = \exp(AL/D) = \exp[(zFu\psi/L)(L/D)] = \exp(zFu\psi/uRT)$$

$$= \exp(zF\psi/RT) \tag{9.95}$$

the equation can be written as

$$J = -A\{[c^o - c^L \exp(zF\psi/RT)]/[1 - \exp(zF\psi/RT)]\} \tag{9.96}$$

The leading coefficient A can also be simplified using the Einstein relation,

$$A = zFu\psi/L = zF(D/RT)(\psi/L) = (zF\psi/RT)(D/L) \tag{9.97}$$

The first factor is just the ratio of the electrical energy to the thermal energy while the second factor is the permeability, P. The equation is

$$J = -P\left(\frac{zF\psi}{RT}\right) \frac{[c^o - c^L \exp(zF\psi/RT)]}{[1 - \exp(zF\psi/RT)]} \tag{9.98}$$

Although this equation is complicated, a number of limiting cases clarify its physical interpretation. For example, a large positive applied potential will accentuate the terms containing the exponential, and the equation reduces to

$$J \approx -P(zF\psi/RT)\{[-c^L \exp(zF\psi/RT)]/[-\exp(zF\psi/RT)]\} \quad (9.99)$$

$$= -P(zF\psi/RT)(-c^L) = +(PzFc^L/RT)\psi \quad (9.100)$$

Using the equation $P = D/L$, this equation can be restated in terms of the electric field and the charge flux:

$$I = -(z^2F^2Dc^L/RT)\,E \quad (9.101)$$

The conductivity is then

$$\kappa = (z^2F^2Dc^L/RT) \quad (9.102)$$

For the electrodiffusion of a cation with a large positive potential, the conductivity includes c^L, because most of the charge carriers will be moving from this bath under the applied potential. For large negative potentials, the nonexponential terms dominate for cation flow and the conductivity is

$$\kappa = (z^2F^2Dc^o/RT) \quad (9.103)$$

The conductivities in the two opposite directions differ, and a positive potential can produce a different current than an equal negative potential when the concentrations of the baths differ.

Another limiting case arises as the potential across the membrane approaches zero. If a potential $\psi = 0$, is substituted into Eq. (9.96), the denominator is zero and the flux is indeterminate. To determine this flux, the flux near zero must be determined and be allowed to approach zero potential as a limit.

For small potentials, the exponentials are expanded to their linear term,

$$\exp(ZF\psi/RT) = 1 + (zF\psi/RT) \quad (9.104)$$

and the flux equation is

$$J = -(PzF\psi/RT)[c^o - c^L(1 + zF\psi/RT)]/[1 - (1 + zF\psi/RT)] \quad (9.105)$$

$$= -(PzF\psi/RT)(c^o - c^L - c^LzF\psi/RT)/(-zF\psi/RT)$$

$$= P(c^o - c^L - c^LzF\psi/RT) \quad (9.106)$$

In the limit of zero potential, the equation reduces to

$$J = -P(c^L - c^o) \quad (9.107)$$

which is the expression for simple diffusion when no electrical driving force is present.

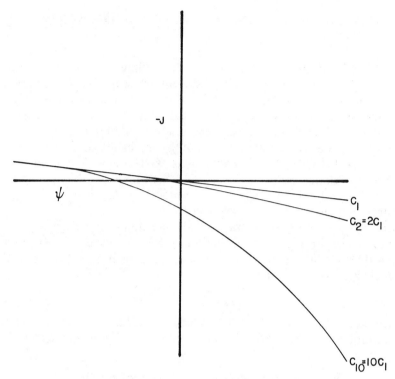

Fig. 9.8   Membrane flux versus potential when Bath 2 concentrations are $1c$, $2c$, and $10c$, using the constant-field equation.

The flux as a function of the transmembrane voltage for a constant c^o and $c^L = c^o$, $2c^o$, and $10c^o$ is shown in Fig. 9.8. For large negative potentials, the flux-versus-voltage curve is linear. Similarly, for large positive potentials, the limiting curve is linear but the slope of the line will be larger for the larger concentrations. When both concentrations are equal, a potential can drive the same number of ions through the membrane in either direction so the current versus voltage curve will be linear over the full range. When the bath concentrations are different, the linear curves will slope in the intermediate region to permit a continuous evolution between the two linear limits as shown.

When the concentration of Bath 2 is 10 times that of Bath 1, the resultant curve of current versus voltage has a shape characteristic of a rectifier. The rectifier passes appreciable current for a positive potential and significantly less for the negative potential. In this model, the "rectification" is controlled by the numbers of ions available in each bath to carry the charge across the membrane.

8. THE GOLDMAN–HODGKIN–KATZ EQUATION

The Goldman equation described the electrodiffusion of a single ion with specific mobility and charge. If additional ions diffuse through the membrane, the net current must include contributions from all these ions. The ions that traverse the membrane most rapidly—that is, are most mobile—will carry the larger portion of the current. The relative mobilities of the ions will play a major role in the determination of the net membrane flux.

Although the Goldman–Hodgkin–Katz (GHK) equation normally describes the relative contributions of K^+, Na^+, and Cl^- ions to the total charge flux, only K^+ and Cl^- ions are used here, as the extension to an arbitrary number of ions will then be obvious.

All ions are assumed to traverse the membrane independently. This "independence principle" states that the each ion permeates the membrane independently with no interference from other ions in the membrane. The total charge flux is written as

$$I = z_K F J_K + z_{Cl} F J_{Cl} \qquad (9.108)$$

where z_K, the K^+ ion charge, is $+1$ and $z_{Cl} = -1$. Each of the fluxes is given by the Goldman equation for that ion with the appropriate parameters. The currents are

$$I_K = \frac{u_K z_K^2 F^2 \psi}{L} \left[\frac{c_K^o - c_K^L \exp(z_K F \psi / RT)}{1 - \exp(z_K F \psi / RT)} \right] \qquad (9.109)$$

and

$$I_{Cl} = \frac{u_{Cl} z_{Cl}^2 F^2 \psi}{L} \left[\frac{c_{Cl}^o - c_{Cl}^L \exp(z_{Cl} F \psi / RT)}{1 - \exp(z_{Cl} F \psi / RT)} \right]$$

Because of the z^2 in the leading term for each current, both terms will add. The effect of ion charge becomes significant in the exponential terms, because the exponentials for the current term will have opposite signs. To permit combination of the two terms, the numerator and denominator for I_{Cl} are each multiplied by

$$\exp(-z_{Cl} F \psi / RT) = \exp(F \psi / RT) = y \qquad (9.110)$$

The two currents are now

$$I_K = u_K (F^2 \psi / L)(c_K^o - c_K^L y)/(1 - y)$$
$$I_{Cl} = u_{Cl}(F^2 \psi / L)(c_{Cl}^o y - c_{Cl}^L)/(y - 1)$$
$$= u_{Cl}(F^2 \psi / L)(c_{Cl}^L - c_{Cl}^o y)/(1 - y) \qquad (9.111)$$

The total current is

$$I_t = (F^2\psi/L)(1 - y)^{-1}[(u_K c_K^o + u_{Cl} c_{Cl}^L) - y(u_K c_K^L + u_{Cl} c_{Cl}^o)] \quad (9.112)$$

The extension of this equation for additional ions is straightforward. Cation concentrations are multiplied by their respective mobilities and added when they describe Bath 1 (c^o); their c^L are multiplied by y and their appropriate u and added. The c^L for each anion is multiplied by the proper u and added; c^o for each anion is multiplied by y and u and added. When $z \neq 1$, this factor must be included. The term y becomes y^z when $z \neq 1$.

Although the GHK equation is useful for determining steady fluxes through the membrane, some knowledge of the mobilities of the ions in the membrane is still required. The form of the equation simplifies at equilibrium when $I = 0$. For example, the equation for a single ion at equilibrium is

$$0 = (uF^2\psi/L)(c^o - c^L y)/(1 - y) \quad (9.113)$$

$$c^o = c^L y = c^L \exp(F\psi/RT) \quad (9.114)$$

or

$$\psi = (RT/zF) \ln(c^o/c^L) \quad (9.115)$$

This is the equilibrium potential.

When the equilibrium condition is applied to the multi-ion GHK equation, the total expression for Na^+, K^+, and Cl^- at equilibrium becomes

$$(u_K c_K^o + u_{Na} c_{Na}^o + u_{Cl} c_{Cl}^L) - \exp(F\psi/RT)(c_K^L u_K + c_{Na}^L u_{Na} + c_{Cl}^o u_{Cl}) = 0$$

$$(9.116)$$

$$\exp(F\psi/RT) = (c_K^o u_K + c_{Na}^o u_{Na} + c_{Cl}^L u_{Cl})/(c_K^L u_K + c_{Na}^L u_{Na} + c_{Cl}^o u_{Cl})$$

$$(9.117)$$

The electrochemical potential is then

$$\psi = (RT/zF) \ln[(c_K^o u_K + c_{Na}^o u_{Na} + c_{Cl}^L u_{Cl})/(c_K^L u_K + c_{Na}^L u_{Na} + c_{Cl}^o u_{Cl})]$$

$$(9.118)$$

At equilibrium, the GHK equation is very similar to the equation for the electrochemical potential of a single ion. The ions are added, but they must be weighted by their respective mobilities. The c^o cationic terms appear in the numerator with the c^L anionic terms. The equation permits resolution of a question that was raised in Chapter 4: how do we determine the electrical potential for a multi-ion system? It is now apparent that we must sum the ions. However, they cannot be summed with equal weight; the faster ions play a larger role in the determination of the equilibrium potential. The mobility is a microscopic parameter of the ion, and thermodynamics does not generally include such parameters. However, thermodynamics does allow corrected concentrations (activities). The thermodynamicist would observe an experimental discrepancy; he would modify the concentrations to obtain the proper correlation. The kinetic study provides additional information; it indicates that the contribution of each ion is related to its ability to permeate the membrane.

The GHK equation was derived from the constant-field Goldman equation. However, in special cases, the equation can be derived without reference to the nature of the field within the membrane. Sandblom and Eisenman developed the GHK equation without invoking the constant-field assumption for a two-cation system. Consider the Nernst–Planck flux equations for K^+ and Na^+:

$$J_K = -u_K c_K [(RT/c_K)(dc_K/dx) + F\, d\psi/dx] \tag{9.119}$$

$$J_{Na} = -u_{Na} c_{Na} [(RT/c_{Na})(dc_{Na}/dx) + F\, d\psi/dx] \tag{9.120}$$

At equilibrium, the sum of these fluxes is zero,

$$J_K + J_{Na} = 0$$

and the pair of Nernst–Planck equations can be written in the form

$$(d/dx)(u_K c_K + u_{Na} c_{Na}) + (F/RT)(u_K c_K + u_{Na} c_{Na})\, d\psi/dx = 0 \tag{9.121}$$

If both sides of this equation are now divided by $u_K c_K + u_{Na} c_{Na}$, the equation becomes

$$d[\ln(u_K c_K + u_{Na} c_{Na})]/dx + (F/RT)\, d\psi/dx = 0 \tag{9.122}$$

which is now integrated between c^o and c^L to give

$$\ln(u_K c_K + u_{Na} c_{Na}) \Big|_0^L = -(F/RT)\,\psi \Big|_{\psi^o}^{\psi^L} \tag{9.123}$$

and

$$\psi = \psi^L - \psi^o = (RT/F) \ln[(u_K c_K^o + u_{Na} c_{Na}^o)/(u_K c_K^L + u_N c_{Na}^L)] \tag{9.124}$$

This result is identical to the GHK expression developed using the constant-field assumption. No assumptions about the field within the membrane were required for this equilibrium solution. We will find a similar situation when we study discrete-state models. In general, the thermodynamic information is found in the numerator, while specifics of the membrane model appear in the denominator and are eliminated at equilibrium.

9. FLUX RATIOS

When the Goldman and GHK equations were limited to zero net flux, a major cancellation of terms occurred and the equations reduced to the electrochemical potential for equilibrium. However, this simplification limits the generality of the equation. An alternative approach develops expressions for the membrane fluxes in opposite directions (the unidirectional fluxes) and forms the ratio of the fluxes. Because many factors in the equation will be common to both of these fluxes, the net equation is far simpler.

When a current flows through the membrane, the nature of the charge carriers (anions or cations) is not known. In addition, the observed current or flux is really a net flux. For example, if 5 particles passed through the cross section of membrane each second, this could be caused by the actual flow of 5 particles. It could also be caused by the flow of 10 particles to the right and the flow of 5 particles to the left in the same time period. The net flux is the difference between unidirectional flows in opposite directions,

$$J_{net} = J_r - J_l \qquad (9.125)$$

where J_r and J_l are the unidirectional flows to the right and left, respectively, as shown in Fig. 9.9. Normally, these unidirectional fluxes cannot be resolved. However, radioactive tracer techniques do permit resolution. Tracer ion is added to one bathing solution, and its loss from this solution or increase in the second solution is monitored by changes in the total observed radioactivity. In this manner, the unidirectional fluxes in both directions can be determined and their ratio can be determined.

To establish the functional form of the ratio, the constant-field approximation is used. However, the expression developed using this equation is extremely general and the constant-field equation provides a vehicle to obtain it.

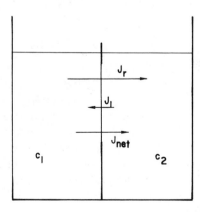

Fig. 9.9 Undirectional fluxes as constituents of a net membrane flux.

Consider two limiting cases of the Goldman equation. When c^o is large relative to c^L, the limiting equation contained only the term in c^o because this bath provided the majority of the charge carriers. All the flow is from left to right, so this limiting case represents an example of a unidirectional flux from left to right,

$$J_r \approx -P(zF\psi/RT)c^o/[1 - \exp(zF\psi/RT)] \qquad (9.126)$$

When c^L is the dominant concentration, the flow will be essentially unidirectional to the left,

$$J_l \approx -P(zF\psi/RT)c^L/[1 - \exp(zF\psi/RT)] \qquad (9.127)$$

For the flux in intermediate regions, Eq. (9.125) is used and reproduces the original Goldman equation. However, we assume that Eq. (9.126) and (9.127) represent the unidirectional flows in this region as well, because the ions flow independently of each other.

A unidirectional flux ratio can be generated using Eq. (9.126) and (9.127),

$$J_r/J_l = (c^o/c^L)\exp(-zF\psi/RT) \qquad (9.128)$$

Detailed information contained in parameters like P is eliminated when the ratio is formed, and the fluxes depend only on the bath concentrations and the transmembrane potential difference. The potential appears in a Boltzmann factor, which simply reflects the difference in energy between the two bathing solutions. The ratio (c^o/c^L) also represents an energy difference between the two baths, because

$$c^o/c^L = K = \exp[-(G^o - G^L)/RT] \qquad (9.129)$$

where the G terms are the chemical free energies for each bath.

Equation (9.128) also shows that the unidirectional flows can be controlled by the magnitude of the applied potential. It is convenient to reexpress the equation in logarithmic form,

$$\ln(J_r/J_l) = \ln(c^o/c^L) - zF\psi/RT \qquad (9.130)$$

A plot of the logarithm of the flux ratio against potential is linear with a slope of $(-zF/RT)$, and this serves as an excellent check of the flux-ratio relation. The flux-ratio relation carries the implicit assumption that all ions traverse the membrane independently.

Hodgkin and Keynes (1955) performed radioactive tracer experiments of K^+ ion flows in squid axon membrane and determined the unidirectional flux ratios. The resultant logarithmic plot [Eq. (9.130)] had a slope that was 2.5 times as large as the predicted slope, as shown in Fig. 9.10. These data suggest a significant lack of independence for ions in the membrane. To

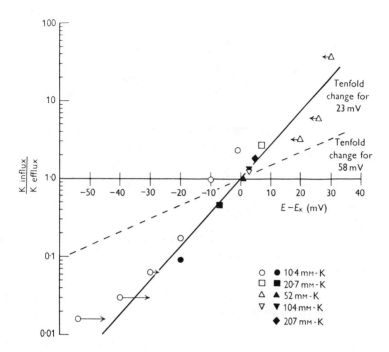

Fig. 9.10 Effect of driving force, $E - E_K$, on potassium flux ratio in fibers poisoned with 0.2 mM DNP (dinitrophenol). [Reproduced with kind permission of Keynes (1955).]

include such nonindependence, a model in which the ions traversed the channel through long, narrow pores was postulated. As the ions traversed the pore, they were likely to encounter other ions, which produced the anomalous behavior. The slope suggested an average of 2.5 interfering ions within the channel.

Hodgkin and Keynes (1955) introduced a mechanical analog to mimic the behavior of the long pores. They constructed two containers that could be connected by either a short or long pore as shown in Fig. 9.11. Each container held balls of different color. For a short channel with 100 balls on the left and 50 balls on the right, the flux to the right was observed to be 2.7 times that of the flux to the left. This is close to the predicted value of 2. When a long, narrow pore replaced the short pore, 18 times as many balls migrated to the right. The long pore contained an average of about 3 balls, and 4 collisions were required to propel a ball through the pore. Because the left container contains twice as many balls, the chance that 4 sequential collisions move a particle through the pore to the right is $(2/1)^4 = 16$ times as likely as 4 sequential collisions to the left. The observed factor of 18 compares favorably with this result.

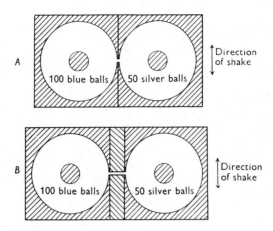

Fig. 9.11 Diagram of mechanical model. (a) Two flat compartments were separated by a narrow gap. (b) The gap width was increased by spacers. [Reproduced with kind permission of Keynes (1955).]

10. DISCRETE STATE DIFFUSION MODELS

The diffusion models determined from the diffusion equation depend on the continuous variable, x. This suggests that the particles move in increments of magnitude dx, so that their motion in the membrane constitutes a continuous flow. Such models work extremely well when the motions of the ions are small relative to the total distance to be traversed. The situation becomes more complicated at the molecular level when the total distance to be traveled is not significantly larger than the ion sizes. For example, a bilayer of 60 Å width is only an order of magnitude larger than the ionic diameters, and its molecular constituents will not be "blurred" into a continuum. A more realistic picture suggests that the ion will traverse some mean free path within the membrane before encountering or colliding with a membrane molecule. For example, a K^+ ion might proceed through a channel in an average of three steps, with each "step" characterized as an encounter with the membrane molecules. As the number of such steps increased, the diffusion models and discrete state models would coalesce.

Since diffusion via a series of discrete jumps is consistent with an intuitive physical picture for ion transport through membranes, we will formulate such a model for comparison with continuum models. Once again, motions of the ions parallel to the surface will be ignored and the problem is one dimensional. The appropriate steady-state model was developed by Parlin and Eyring (1954). The diffusion process is a multistep kinetic process. The

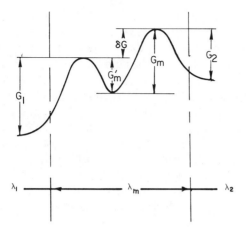

Fig. 9.12 A discrete state transition state model with two barriers. The free energies describe the barrier heights, while the λ_i terms define the spatial ranges for the bathing solutions and membrane.

ion surmounts an energy barrier to enter the membrane and then falls to a metastable state within the membrane, as shown in Fig. 9.12; it can cross an arbitrary number of such barriers with different heights, as shown in Fig. 9.12, before emerging into the bathing solution on the opposite side. This energy profile for the ion path should not be confused with the actual path that the ion takes in the membrane. The ion moves straight from site to site. The energy peaks simply provide a measure of the difficulty and time for each step. The ion will tend toward the state of lowest energy.

To understand the Parlin–Eyring model, an expression for a single ion flux is developed for a model having a single membrane state. In Fig. 9.12, the ion crosses an energy barrier from Bath 1 to the membrane state; it can then cross a second barrier and enter Bath 2. In order to surmount the first barrier from the right, a free energy of G_1 is required. The free energy includes both enthalpic and entropic changes associated with the transition from solution to membrane. To cross the second barrier from left to right, G_m is required. However, the particle can also jump back to Bath 1. The free energy can differ from G_m as shown, and is defined as G'_m. Transitions in both directions must be included.

Particles may also leave Bath 2 and enter the membrane by surmounting a barrier of energy, G_2. Four distinct free energies are thus required for the two-barrier model. The heights of the energy barriers can be used to determine rate constants for the transitions using Eyring absolute rate theory, which postulates rate constants of the form

$$k = (k_B T/h) \exp(-G/RT) \qquad (9.131)$$

where k_B is Boltzmann's constant (1.6×10^{-23} J/K · molecule) and h is Planck's constant (6.6×10^{-34} J · s). The ratio ($k_B T/h$) is approximately 10^{13} s^{-1} at room temperature. If there were no free energy barrier, the time to cross the energy barrier would be approximately one molecular vibration.

The rate constant has the units of s^{-1}. The rate at which ions crossed the barrier would normally be expressed as concentration per second. However, we are interested in a flux rather than a concentration, and the flux units are moles per area per second. For consistency, one distance unit must be eliminated from the rate units. The membrane system is broken into three distinct regions, the membrane and two bathing solutions. If Bath 1 has a concentration of c_1 and we are considering only a unit cross-sectional area of membrane, we assume that all ions in the bath for a distance λ_1 are capable of entering the membrane in a single jump, no matter how far from the membrane they are. The product

$$\lambda_1 k_1 c_1 \tag{9.132}$$

has the units of flux and describes the motion from Bath 1 to the membrane. The expression selects a unit cross section of solution and membrane.

The forward membrane flux over the second barrier is

$$k_m \lambda_m c_m \tag{9.133}$$

and the reverse membrane flux over the first barrier is

$$k_{-m} \lambda_m c_m \tag{9.134}$$

The net flux over the first barrier is then

$$J_1 = k_1 \lambda_1 c_1 - k_{-m} \lambda_m c_m \tag{9.135}$$

while the net flux for the second barrier is

$$J_2 = k_m \lambda_m c_m - k_2 \lambda_2 c_2 \tag{9.136}$$

These flux equations contain a number of unknowns, including the actual concentration, c_m, of ions in the membrane. If this variable could be eliminated, the final solution would contain only the bath concentrations.

This concentration can be eliminated if the system is at a steady state because the number of ions within the membrane must remain constant. To satisfy this condition, the flux over the first barrier must equal the flux over the second barrier. For example, if 10 particles are entering the membrane region, another 10 particles must be leaving the region. Thus,

$$J = J_1 = J_2 \tag{9.137}$$

and the pair of equations can be combined to eliminate c_m. The second flux equation is solved for $\lambda_m c_m$,

$$\lambda_m c_m = [J + (k_2 \lambda_2 c_2)]/k_m \tag{9.138}$$

where the membrane width and concentration are paired because they both contain information on the membrane. Equation (9.138) is now substituted into Eq. (9.135) to eliminate $\lambda_m c_m$,

$$J = k_1 \lambda_1 c_1 - (k_{-m}/k_m)[J + k_2 \lambda_2 c_2] \tag{9.139}$$

The terms containing J are now combined to permit solution for J,

$$J[1 + (k_{-m}/k_m)] = k_1 \lambda_1 c_1 - (k_{-m}/k_m)k_2 \lambda_2 c_2 \tag{9.140}$$

The final result is

$$J = [k_1 \lambda_1 c_1 - (k_{-m}/k_m)k_2 \lambda_3 c_2]/[1 + (k_{-m}/k_m)] \tag{9.141}$$

Although this expression for the flux shows the direct dependence on the individual rate constants of the system, these rate constants are still unknowns and the expression is still quite complicated. We can gain some additional insights if Eyring rate theory is used to replace the rate constants with expressions for the barrier free energies. To illustrate this change, the ratio (k_{-m}/k_m) is converted to an energy-equivalent form. Using Eyring rate expressions for each k,

$$k_{-m}/k_m = [(k_B T/h) \exp(-G_{-m}/RT)]/[(k_B T/h) \exp(-G_m/RT)] \tag{9.142}$$

$$= \exp[-(G_{-m} - G_m)/RT] \tag{9.143}$$

The ratio is related to the difference in the heights of the two barriers and is shown as δG in Fig. 9.12. Note that a large barrier to reverse reaction (large G_m) will decrease the ratio of rate constants and enhance the flow to Side 2.

The numerator can also be recast with free energies; the equation is rearranged to

$$k_1 \lambda_1 [c_1 - (k_{-m}/k_m)(k_2/k_1)(\lambda_2/\lambda_1)c_2) \tag{9.144}$$

and

$$\lambda = \lambda_1 = \lambda_2 \tag{9.145}$$

When the Eyring expressions with their appropriate free energies are substituted the expression becomes

$$k_1 \lambda [c_1 - \exp(-G_{-m}/RT) \exp(+G_m/RT)\exp(-G_2/RT) \exp(+G_1/RT)c_2] \tag{9.146}$$

and the collective free energy is

$$G_2 - [G_m - G_{-m}] - G_1 = G_2 - \delta G - G_1 \tag{9.147}$$

This combination of free energies is illustrated in Fig. 9.12, where G_2 is the height from Bath 2 to the peak of the second barrier, and $G_1 + \delta G$ is the height from Bath 1 to this barrier. The difference between them is the free energy difference between Bath 2 and Bath 1. This free energy difference is $+\Delta G$, and the flux expression is

$$J = k_1\lambda\{c_1 - \exp[(-\Delta G/RT)c_2]\}/[1 + \exp(\delta G/RT)] \qquad (9.148)$$

The free energy can be related to the electrical potential applied to the membrane via the relation

$$\Delta G = -zF\psi \qquad (9.149)$$

When this expression is substituted into the equation, it becomes

$$J = k_1\lambda\{c_1 - c_2\exp[(zF\psi/RT)]\}/[1 + \exp(\delta G/RT)] \qquad (9.150)$$

The numerator is identical to that of the Goldman equation. The equilibrium information is contained in this portion of the expression and is expected to be present in all models.

The leading coefficient of these equation, $k_1\lambda$, has units of distance per second and is thus the discrete state version of the permeability coefficient

$$P = k_1\lambda \qquad (9.151)$$

In order to develop the discrete state diffusion coefficient, an additional distance parameter must be included. This is λ_m, the membrane thickness,

$$D = k_1\lambda_1\lambda_m \qquad (9.152)$$

Although Eq. (9.150) was derived for a single-membrane state, the extension to multiple-membrane states is simple. The numerator, describing the equilibrium information, is unchanged. There are two terms in the denominator for the single-membrane state model representing the two barriers. In a multistate system, the ith barrier is described by a free energy, δG_i, describing the height of this barrier relative to the first barrier. The expression for n barriers is

$$J = k_1\lambda[c_1 - c_2\exp(-\Delta G/RT)]\bigg/\bigg[1 + \sum_i^{n-1} \exp(\delta G_i/RT)\bigg] \qquad (9.153)$$

11. THE GOLDMAN AND PARLIN–EYRING MODELS

The Parlin–Eyring and Goldman models yielded identical forms for the the numerator, as this portion of each equation contained the equilibrium information. The denominators, with their specific information on the model, are quite different. The Goldman produces a difference of two terms, while

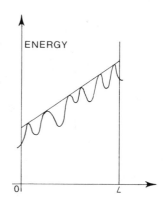

Fig. 9.13 Relation between the Goldman and Parlin–Eyring models. The discrete barrier heights must increase linearly across the membrane.

the Parlin–Eyring is a sum of terms. However, with the proper choice and number of energy barriers, the Parlin–Eyring model reduces to the Goldman equation.

Consider the situation where the barrier heights increase linearly across the membrane, as shown in Fig. 9.13. The membrane potential at each barrier is proportional to the distance into the membrane, and the δG_i terms are

$$\delta G_i = -(x_i \Delta G/LRT) \qquad (9.154)$$

where x_i/L represents the fraction of free energy utilized to that distance in the membrane. For the electrical potential difference, the expression is

$$+zF\psi x_i/LRT \qquad (9.155)$$

If the number of such barriers is now increased, the total denominator summation (including the 1) can be replaced by an integral

$$\int_0^L \exp(xzF\psi/LRT)\,dx = \int_0^L \alpha^x\,dx \qquad (9.156)$$

where

$$\alpha = \exp(zF\psi/LRT) \qquad (9.157)$$

The integral is solved using the identity

$$\alpha^x = \exp(x \ln \alpha) \qquad (9.158)$$

to give the result

$$\int_0^L \alpha^x\,dx = (\alpha^L - 1)/\ln(\alpha) = [\exp(zF\psi/RT) - 1]/(zF\psi/LRT) \qquad (9.159)$$

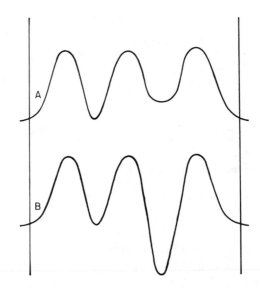

Fig. 9.14   Two barrier models that produce equal net fluxes.

When this expression replaces the denominator in the Parlin–Eyring equation, the Goldman equation results:

$$J = -\left(\frac{zF\psi}{LRT}\right)k_1\lambda\frac{[c_1 - c_2\exp(zF\psi/RT)]}{1 - \exp(zF\psi/RT)} \qquad (9.160)$$

A second limiting case of some interest is also used to simplify the Parlin–Eyring model. If all the barriers have equal heights, then all the exponentials in the sum become unity and the denominator is equal to the total number of barriers n. This result is particularly interesting because the result is independent of the depth of the wells. Figure 9.14 shows two very different energy-barrier diagrams that will produce exactly the same flux. Although the deeper wells might be expected to produce a slower rate of flow, the depth will be compensated by a large ion concentration in the wells. The rate over the barrier, which is a function of both the depth and the concentration in the well, will be the same for both models.

PROBLEMS

1.   Write the equation of continuity for a spherical surface.

2.   Estimate the diffusion coefficient for a particle that takes jumps of 5 Å every 10^{-9} s.

3. Determine the plot for the concentration at the center of a membrane as it decays to equilibrium. The initial concentrations on each side are 0.05 M, and Side 2 is raised to 0.1 M at $t = 0$. Determine the time required for this concentration to drop to $1/e$ of its initial value.

4. A pressure gradient can be used to generate a transmembrane force. Develop a flux equation containing a concentration gradient and a pressure gradient.

5. Fick's second law is generated from Fick's first law using the equation of continuity. Generalize the Nernst–Planck equation to the time-dependent form.

6. Develop GHK equation for $z:z$ electrolyte and a $z:1$ electrolyte.

7. Show that permeability coefficients P can be used in place of mobilities in the GHK equation.

8. In an excitable membrane, the flux ratios of inward and outward currents for the K^+ ion determined with radioactive tracers are found to be ($RT/F = 25\,\mathrm{mV}$):

$V(\mathrm{mV})$	$i_e/i_i = J_r/J_1$
-50	14.9
-25	5.5
0	2.0

What is the K^+ concentration ratio? Are the fluxes independent? Explain.

9. Find an expression for the equilibrium potential generated by two $CaSO_4$ solutions (c_1 and c_2) separated by a membrane permeable to both ions. The permeability coefficients are P_{Ca} and P_{SO_4}.

10. The Goldman equation was derived for cations. Show the form of the equation expected for anions.

11. Determine the following limiting cases for the Goldman equation: (a) large positive voltages; (b) $c_1 \ll c_2$; (c) equal concentrations; (d) zero flux.

12. Determine the membrane flux using the Parlin–Eyring model for a membrane that contains N barriers of equal height.

13. Using the Goldman equation, develop unidirectional flux ratios when two univalent species of like charge contribute to the flow. Show the form of the flux ratio.

14. Using the case for equal barrier heights in problem 12, determine the fluxes for two- and three-barrier models. (a) Calculate the conductances for each model. (b) Develop an expression to show how the conductance changes with increasing numbers of barriers.

Chapter 10 _____
Irreversible Thermodynamics

1. REVERSIBLE AND IRREVERSIBLE THERMODYNAMICS

Reversible thermodynamics introduces a variety of work terms, which are generally characterized as the product of an intensive and an extensive thermodynamic parameter; this product has the units of energy. The transition from reversible to irreversible thermodynamics can be illustrated by considering one pair of parameters that are used extensively in both their reversible and irreversible forms. These parameters will then serve as the basis for a general discussion of irreversible thermodynamic parameters in Section 2.

The thermodynamic parameters of interest are the potential ψ (the intensive variable) and the amount of charge moved through this potential, dq. The energy is

$$\psi dq \tag{10.1}$$

The differential form of this expression is significant, because it suggests that an infinitesimal amount of charge flows and the potential is not perturbed by this charge flow. If a large amount of charge flows through a potential difference, such as current in a resistor, then energy is dissipated as heat in the resistor. For such currents, the process is irreversible. This suggests that Ohm's law of the form

$$i = G\psi \tag{10.2}$$

provides some measure of the irreversible behavior of the system. To establish the connection, Ohm's law must be expressed in a more general format. When we introduced the conductivity κ in Chapter 7, we defined the total current that crossed a unit area of material and moved the charge for some unit distance across the material; that is, we observed a charge flux characterized by the units

$$\text{charge/unit area} \cdot \text{time} \tag{10.3}$$

In addition, the potential difference appeared across the unit distance in a homogeneous material, so that the electric field is

$$\text{volts/unit distance} \tag{10.4}$$

for this unit cube. Ohm's law is written as

$$I = \kappa E \tag{10.5}$$

where I is the charge flux and E is the electric field.

Ohm's law describes the irreversible flow of current, so it is worth comparing its variables with those for the reversible thermodynamic expression, Eq. (10.1). The charge dq is replaced with a rate of current flow,

$$I = (1/A)\,dq/dt = i/A \tag{10.6}$$

The concept of rate C/s never enters reversible thermodynamics. Thermodynamics determines the energy required to move a test charge, and not the time that might be required. Current determines the dissipation of energy in the resistor, that is, the rate of energy production.

Ohm's law states that the rate of current flow will depend on the electric field rather than on the potential. The field is a measure of the force experienced by the charge. When the charged particle experiences a large force, the charge motion will be highly irreversible. Consider a potential of 0.1 V for two different situations. The 0.1 V is applied to two plates separated by 1 mm, or it is applied to a membrane of 10 nm thickness. The resultant electric fields are 100 V/m and 10^7 V/m, respectively. A charged particle will accelerate dramatically in the membrane field. As this accelerating particle encounters other particles in the medium, it will dissipate its energy as heat. Although the 1 mm separated plates have the same potential, particle acceleration and energy dissipation are small because of the smaller electric field. Thus, it is a force, here expressed as the electric field, that dictates how irreversible the system is.

A field can always be expressed as the derivative of a potential. For a one-dimensional system,

$$E = -d\psi/dx \tag{10.7}$$

This could be generalized to a three-dimensional system by using the gradient operator,

$$\mathbf{E} = -\nabla\psi \tag{10.8}$$

where $\mathbf{E}$ is now a vector in three-dimensional space.

Similar expressions result for any potential energy. The derivative of this energy with respect to a specific directional variable gives the force in the direction of that variable,

$$F_x = -dV/dx \qquad (10.9)$$

or

$$F = -\nabla V \qquad (10.10)$$

In irreversible thermodynamics, the forces and the fluxes of material produced by these forces are the central variables.

The formal definitions of field and force introduce a negative sign, and this suggests that Ohm's law should be written as

$$I = -\kappa E \qquad (10.11)$$

The sign relates the direction of flow and the direction of the applied field. As discussed in Chapter 9, the charged positive test particle moves from the higher potential to the negative potential. It also moves in the direction of the field, so the potential and the field must be opposites. The minus sign expresses this fact.

2. GENERALIZED FORCES AND FLUXES

The thermodynamic charge flux and its associated driving field were developed directly from the conjugate thermodynamic variables ψ and dq. In Chapter 9, a similar analysis was used to develop Fick's law of diffusion. The flux of particles J described by Fick's law was produced by the gradient of the chemical potential,

$$F = -d\mu/dx \qquad (10.12)$$

The parallel is clear. The force was the negative gradient of a potential. The resultant flow of particles under the concentration gradient gave rise to a flux of the form

$$J = (1/A) \, dn/dt \qquad (10.13)$$

where dn/dt describes the number of moles of particles flowing across some unit cross section per second. The force and flux were derived from the intensive chemical potential μ and its extensive conjugate dn.

These two examples suggest a generalization to any arbitrary pair of conjugate thermodynamic variables. Each conjugate pair can be expressed as the product of a general intensive parameter designated V and a general extensive parameter designated as Q, so that the energy for this conjugate pair is

$$\text{energy} = V \, dQ \qquad (10.14)$$

For example, for $P \, dV$ work, pressure becomes the intensive potential V and the intensive parameter dV becomes the generalized "charge." For a surface tension energy of the form

$$\text{energy} = \gamma \, dA \qquad (10.15)$$

γ, the intensive energy per unit area, becomes V and dA is dQ.

A generalized force X is now defined as

$$X = -dV/dx \qquad (10.16)$$

for one dimension or

$$\mathbf{X} = -\nabla V \qquad (10.17)$$

for three dimensions. The conjugate flux of generalized charge is now

$$J = (1/A) \, dQ/dt \qquad (10.18)$$

By analogy with Ohm's law, the flux of generalized charge will be proportional to the magnitude of the generalized force,

$$J = -LX \qquad (10.19)$$

where L is the proportionality constant. For Fick's law, L is the diffusion constant D, while L is the conductivity κ for Ohm's law.

To illustrate utilization of these generalized forces and fluxes, consider the irreversible thermodynamics which arise in a study of pressure volume work, $P \, dV$. The force is now

$$X = -dP/dx \qquad (10.20)$$

and the flux is

$$J = (1/A) \, dV/dt \qquad (10.21)$$

In a reversible thermodynamic expansion, the internal and external pressures on the piston are infinitesimally close to each other and the piston must move at an infinitesimal rate. If we break the volume both inside and outside the cylinder into a set of equal cells, then the expansion is equivalent to moving some volume cells from the outside to the inside of the cylinder. In other words, we describe a flux of volume cells across the boundary to the cylinder interior. The flux of volume cells is described by Eq. (10.21). For a piston expanding in only one direction with cross-sectional area A, Eq. (10.21) simplifies to

$$J = dx/dt \tag{10.22}$$

where x is the variable in the direction of the piston. In other words, the flux is simply the velocity of the piston. The largest velocities are expected when the expansion is least reversible.

Equation (10.20) for the pressure gradient emphasizes a fact often overlooked in conventional thermodynamics: the internal and external pressures must be different for the piston to move. The pressure gradient will occur across the piston width in this case. Because this is constant, the magnitude of the gradient is dictated by the actual pressure difference. A large pressure difference will produce a large transfer of volume cells. The parallel to Ohm's law is now

$$dx/dt = L\Delta P/a \tag{10.23}$$

where a is the piston thickness.

The $T\,dS$ energy term can also be extended to an irreversible linear relationship. In this case, a temperature difference between the cell and its surroundings would produce the force

$$X = -dT/dx \tag{10.24}$$

while the flow of entropy into the system would describe the flux,

$$J_s = (1/A)\,dS/dt \tag{10.25}$$

Because

$$dS = q_{rev}/T \tag{10.26}$$

the equation is changed to

$$J_s = (1/AT)\,dq_{rev}/dt \tag{10.27}$$

Because the influx of heat and entropy are intimately related, we now note that we have developed Fourier's law in which a temperature gradient gives rise to a heat flux,

$$J_q = -\lambda\,dT/dx \tag{10.28}$$

Although we can include the flux of entropy into the system as one additional generalized charge flux, entropy and heat share a crucial position because the irreversible processes produced by the various works, such as $P\,dV$ or $\psi\,dq$, will result in energy dissipation: that is, they will add energy to the system. Thus, in irreversible thermodynamics, we will focus on the energy produced by the different work terms as they affect the rate of heat production or entropy production in the system.

3. THE DISSIPATION FUNCTION

When current flows through a resistor, heat is generated in that resistor, and it is this heat that becomes interesting in irreversible thermodynamics. The parameter of interest is not the total heat but the rate of heating, that is, the power. For a resistor that produces a current i when a potential of ψ is applied, the power P is

$$P = i\psi \qquad (10.29)$$

This expression must be converted to one utilizing our formal definitions for field and charge flux, so that

$$P = IE \qquad (10.30)$$

is the power which would be produced in a unit volume of resistor. This is verified by inserting the units,

$$P = I \quad (\text{charge}/\text{m}^2 \cdot \text{s})E(\text{J}/\text{charge} \cdot \text{m}) \qquad (10.31)$$

$$= IE \quad (\text{J}/\text{m}^3 \cdot \text{s}) \qquad (10.32)$$

The use of our general forces and fluxes then gives a power per unit volume.

$$P = [(1/A) \, dQ/dt](dV/dx) = JX \qquad (10.33)$$

For a system involving several conjugate energy pairs, the total dissipated power is simply the sum of products of conjugate forces and fluxes,

$$P = \sum J_i X_i \qquad (10.34)$$

The power generated in the unit volume due to all the work terms is all converted to heat in the system. Thus, Eq. (10.34) might also be written as

$$dq'/dt = \sum J_i X_i \qquad (10.35)$$

where q' is the heat per unit volume. Because q or q' are not standard thermodynamic variables, the expression is often rewritten in terms of the entropy,

$$dq'/dt = T \, dS/dt = \sum J_i X_i \qquad (10.36)$$

The product $T \, dS/dt$ is called the dissipation function, represented by the Greek letter Φ,

$$\Phi = T \, dS/dt \qquad (10.37)$$

The entropy in these expressions is an entropy per unit volume.

The definition of dq'/dt as $T \, dS/dt$ presents an additional problem of interest. Temperature differences will give rise to the flow of entropy, and this suggests that T is also involved in some differential manner. In fact, it is actually involved in two distinct ways. A temperature gradient will induce a flow of heat or entropy across the boundary into the system. However, such heat flows should also produce temperature gradients within the system. When such gradients decay so that the system becomes more random, the entropy within the system should rise. There are thus two sources of entropy associated with the flow of heat.

To develop expressions for the two entropy forms, we will consider flows for a one-dimensional system. Our results are then easily generalized to three-dimensional systems.

We assume that the flow of heat into the system is the only way the system energy can be increased — that is, that there are no work terms. The rate of heat input can then be equated to the rate of change of the internal energy,

$$dU/dt = T\, dS/dt \qquad (10.38)$$

where U is the internal energy per unit volume. This expression can be solved for the rate of entropy change,

$$dS/dt = (1/T)\, dU/dt \qquad (10.39)$$

Because both the temperature gradients and the rate of internal energy change are important, this equation must be modified to include the temperature gradients. This is accomplished by invoking the conservation of energy. Energy that crosses the system boundary must be present in the system. This is expressed by the equation of continuity for energy,

$$\partial U/\partial t = -\partial J_u/\partial x \qquad (10.40)$$

where J_u is the flow of internal energy across the boundary. The equation now permits us to reexpress Eq. (10.37) in terms of gradients,

$$dS/dt = -(1/T)\,(dJ_u/dx) \qquad (10.41)$$

The gradient of temperature can now be introduced by noting

$$\left(\frac{d}{dx}\right)\left(\frac{J_u}{T}\right) = \left(\frac{1}{T}\right)\frac{dJ_u}{dx} + \frac{J_u\, d(1/T)}{dx} \qquad (10.42)$$

Solving for $(1/T)\, dJ_u/dx$ and substituting into Eq. (10.39) gives

$$\frac{dS}{dt} = -\left(\frac{d}{dx}\right)\left(\frac{J_u}{T}\right) - J_u\frac{d(1/T)}{dx} \qquad (10.43)$$

The second term is associated with temperature gradients within the system, induced by the flux J_u. The first term, involving the gradient of J_u/T, is then associated with entropy changes produced by energy flow into the system, because the energy flow over T is characteristic of an entropy flow.

The appearance of internal and external entropy generation terms suggests a modified form of the equation of continuity for the entropy. For a conserved quantity such as energy, the rate of change of internal energy is related entirely to flux across the boundary,

$$dU/dt = -dJ_u/dx \qquad (10.44)$$

For the entropy rate of change, the rate must be balanced by entropy crossing the boundary plus entropy generated spontaneously within the system. The equation of continuity is

$$dS/dt = -dJ_s/dx + \sigma \qquad (10.45)$$

where σ is the rate of internal entropy generation. Comparing Eqs. (10.43) and (10.41) yields

$$J_s = J_u/T \qquad (10.46)$$

and

$$\sigma = \frac{J_u \, d(1/T)}{dx} \qquad (10.47)$$

Equation (10.43) is a scalar equation. In three dimensions it becomes

$$dS/dt = -\mathbf{V} \cdot \mathbf{J}_s + \sigma \qquad (10.48)$$

while Eq. (10.45) becomes

$$\sigma = \mathbf{J}_u \cdot \mathbf{V}(1/T) \qquad (10.49)$$

We have now shown that the rate of change of entropy in an irreversible system can be expressed in two interrelated forms. If temperature gradients are not a central concern — for example, fluxes through membranes in which the temperatures of the two baths are equal — then it is convenient to use the dissipation function $\Phi = T \, dS/dt$. If the temperature gradients are important, then dS/dt is used and the temperature must be moved to the right side of the equation. The difference between the two forms can be illustrated using the chemical potential and the flux of particles. The dissipation function gives the product

$$\Phi = JX = [(1/A) \, dn/dt] \, (d\mu/dx) \qquad (10.50)$$

For dS/dt, the temperature is moved to the right-hand side,

$$dS/dt = [(1/T) \, d\mu/dx]J_n \qquad (10.51)$$

The temperature can now be combined with the chemical potential using some of the techniques developed previously. For example,

$$\mu' = \mu/T \qquad (10.52)$$

can be defined as the chemical potential when variations in temperature are also possible. The resultant force will contain gradients involving both μ and T.

Most of the systems examined in this chapter will not include temperature gradients, and these systems will be formulated in terms of the dissipation function. However, both formats appear extensively. It is usually possible to distinguish between the $T\,dS/dt$ and dS/dt formats, because the latter generally contains the reciprocal T in either its force or flux expression.

4. CHEMICAL REACTIONS

Thermodynamic work expressions have led to force–flux pairs that describe entropy generation within the system. The fluxes involve the motion of some generalized charge across the system boundary. A chemical reaction constitutes a situation in which energy and entropy can be generated entirely within the system. If the reaction is exothermic, heat will be released to the system, and some formalism is required to describe the entropy production. Because the dissipation function reflects power—that is, energy per time— it is apparent that the rate of reaction must now play a role.

Although the chemical reaction has no spatial flux, we can consider the motion from reactants to products as a flux. Because such a process is nondirectional, the flux defined for such a process is called a scalar flux. The scalar flux is selected as the rate of reaction. For the reaction

$$A \longrightarrow B \tag{10.53}$$

the rate is

$$d[B]/dt \tag{10.54}$$

where $[B]$ is the concentration of B. This rate is renamed the degree of advancement with the symbol ξ,

$$d\xi/dt = d[B]/dt \tag{10.55}$$

This definition raises an interesting problem. If the rate had been chosen as the rate of loss of A,

$$-d[A]/dt \tag{10.56}$$

the degree of advancement would have to be negative. Two definitions for ξ are possible.

The problem is exacerbated for reactions of the form

$$aA + bB \longrightarrow cC + dD \tag{10.57}$$

where each of the leading stoichiometric coefficients could be different. Any of four possible rates might be chosen for the degree of advancement.

A common degree of advancement can be proposed for each reaction with a slight modification of our definition. The rate for any species is divided by the stoichiometric coefficient for that species. In addition, products are always assumed to have a positive coefficient while reactants always have a negative coefficient. For Eq. (10.51), the degree of advancement can now be defined in two equivalent ways,

$$d\xi/dt = (1/+1)\,d[B]/dt = (1/-1)\,d[A]/dt \tag{10.58}$$

The degree of advancement now describes both the rate of gain of B and the rate of loss of A.

For Eq. (10.55), the degree of advancement is equal to any one of four normalized rate processes

$$d\xi/dt = -a^{-1}\,d[A]/dt = -b^{-1}\,d[B]/dt = +c^{-1}\,d[C]/dt = +d^{-1}\,d[D]/dt \tag{10.59}$$

To define the force conjugate to this chemical flux, consider the units for the degree of advancement,

$$(1/a)\,d[A]/dt = (1/\text{moles})(\text{moles/liter time}) \tag{10.60}$$

$$= \text{liter}^{-1}\,\text{time}^{-1}$$

The product of the force and the flux must give an energy per unit volume per unit time. To produce such a consistent result, the force used in chemical reactions must have the units of energy. This suggests an obvious definition for the force. The reaction will proceed when its free energy is not equal to zero. A large free energy difference should indicate a very vigorous irreversible reaction, although this does not have to be the case. Thus, we define a chemical affinity A, our reaction force, as the free energy difference for the reaction. This may be expressed in terms of the chemical potentials of the reaction molecules for Eq. (10.51) as

$$A = 1\mu_B - 1\mu_A \tag{10.61}$$

Note that the leading coefficients are the stoichiometric coefficients with the proper signs. For the reaction of Eq. (10.55), the chemical affinity is

$$A = c\mu_C + d\mu_D - a\mu_A - b\mu_B \tag{10.62}$$

The dissipation function for a given chemical reaction is

$$T\,dS/dt = JX = A\xi \tag{10.63}$$

Thus, our definitions for degree of advancement and chemical affinity parallel the definitions that we developed for vector processes involving flow across the boundaries. However, one additional relationship is required to complete the parallel. The vector fluxes obeyed a linear relationship

$$J = -LX \qquad (10.64)$$

This suggests a parallel relationship of the form

$$\xi = -LA^{\backprime} \qquad (10.65)$$

In effect, we are saying that the rate of reaction must be directly proportional to the free energy difference. This is not our usual definition of rate processes. However, this linear relationship will hold when the reaction is close to equilibrium, that is, low chemical affinity.

To demonstrate linear behavior, consider the simple isomerization reaction

$$A \underset{k_r}{\overset{k_f}{\rightleftharpoons}} B \qquad (10.66)$$

with forward and reverse rate constants k_f and k_r. If A and B are gaseous, their chemical potentials are a function of their pressures,

$$\mu_A = \mu_A^\circ + RT \ln P_A \qquad (10.67)$$

and

$$\mu_B = \mu_B^\circ + RT \ln P_B \qquad (10.68)$$

and the chemical affinity for the reaction is

$$A = \mu_B - \mu_A = \mu_B^\circ - \mu_A^\circ + RT \ln P_B - RT \ln P_A \qquad (10.69)$$

$$= \Delta\mu^\circ + RT \ln (P_B/P_A) \qquad (10.70)$$

where P_B and P_A are now the nonequilibrium pressures of the gases. The equilibrium pressures are described in terms of $\Delta\mu^\circ$ using the equilibrium relation

$$\Delta\mu^\circ = -RT \ln K_p \qquad (10.71)$$

where

$$K_P = P_B^\circ / P_A^\circ \qquad (10.72)$$

is the equilibrium constant. This expression can be inserted into Eq. (10.68) to give

$$A = -RT \ln K_p + RT \ln (P_B/P_A)$$

$$= RT \ln [K_p^{-1}(P_B/P_A)] \qquad (10.73)$$

Because we will be interested in the nonequilibrium ratio of pressures, this equation is solved for this ratio,

$$K_p^{-1}(P_B/P_A) = \exp(A/RT) \tag{10.74}$$

This expression with A must now be connected with the degree of advancement. Using the pressures as a measure of concentration, the rate expression is written as

$$d\xi/dt = k_f P_A - k_r P_B \tag{10.75}$$

$$= k_f P_A[1 - (k_r/k_f)(P_B/P_A)] \tag{10.76}$$

The second term in the equation is similar to Eq. (10.72). The two expressions can be equated if we note that the degree of advancement must be zero at equilibrium,

$$k_f P_A^o - k_r P_B^o = 0 \tag{10.77}$$

The equilibrium ratio of pressures K_p is then

$$K_p = P_B^o/P_A^o = k_f/k_r \tag{10.78}$$

and Eq. (10.74) becomes

$$d\xi/dt = k_f P_A[1 - K_p^{-1}(P_B/P_A)] \tag{10.79}$$

Inserting Eq. (10.72) gives

$$d\xi/dt = k_f P_A[1 - \exp(A/RT)] \tag{10.80}$$

as the relationship between the degree of advancement and the chemical affinity. However, the relationship is nonlinear, and the nonlinearity will increase as the system moves further from equilibrium. The expression will become linear only when the system is close to equilibrium as characterized by a small value of A. If $A/RT < 1$, the exponential can be expanded as

$$\exp(A/RT) = 1 - (A/RT) \tag{10.81}$$

to give

$$d\xi/dt = k_f P_A(-A/RT) = -(k_f P_A/RT)A \tag{10.82}$$

This linear relationship holds only near equilibrium, so that the exponential can be expanded and the term P_A remains relatively constant.

5. ELECTROKINETIC EFFECTS

When various potential gradients are applied across membrane systems, the resultant forces will generate fluxes through the membrane. For example, we already know that an applied potential will generate an electric field that stimulates a flow of charge—that is, ions—through the membrane. Ohm's law describes the resultant charge flux

$$I = \kappa E \qquad (10.83)$$

A pressure gradient produces a volume flux, that is, cylinder motion. However, no piston exists in the membrane system. If the membrane is permeable to water, a pressure difference across the membrane will produce a volume flow of water across the membrane. The moles of water that cross the membrane will consitute the flux in a system in which water is the only species present. Alternatively, the flux can be expressed by the total volume of water which crosses the membrane. This alternate definition parallels the development for the moving piston. The flux then has the units

$$(\text{volume/area} \cdot s) = (\text{distance/s}) \qquad (10.84)$$

The volume flux is then related to the velocity at which water molecules cross the membrane. The volume flux J_v and the water flow velocity v can be used interchangeably.

The volume flux is directly proportional to the transmembrane pressure difference,

$$J_v = L_p \Delta P \qquad (10.85)$$

The pressure gradient ΔP is the actual pressure difference divided by the membrane thickness. Because this thickness is constant, it can be incorporated into the constant L_p, which is the filtration coefficient for the membrane.

The equations for Ohm's law and the volume flux are valid for situations in which only ions or only water can flow through the membrane. The situation becomes more complicated when both ions and water can traverse the membrane. For example, an applied pressure difference might create a flow of solution across the membrane. Would the resultant flow of the ions in solution constitute a current?

This question may be answered by reexamining the electroosmotic phenomena discussed in Chapter 8. If the water ions passed through a neutral channel that was large enough to permit the flow of either cations or anions, no net flow of charge would be possible. Each cation would be balanced by an anion. The flow of solution produces a current only when ions of one charge traverse the membrane faster than those of opposite charge. Such differences are generated when the transmembrane channels are lined with charge. The zeta potential produced by the charged walls of the channel produces the differential ion flows, which register as a net membrane current. Thus, an applied pressure difference, which produces a movement of ions relative to the solvent, will generate a current. The streaming current was introduced in Chapter 8 as a linear expression,

$$I = (\varepsilon\zeta/\eta)\Delta P \qquad (10.86)$$

where ε is the dielectric constant,

$$\varepsilon = \varepsilon_r\varepsilon_0 \qquad (10.87)$$

and where ζ is the zeta potential and η is the viscosity coefficient. Equation (10.86) is generated by multiplying the streaming potential [Eq. (8.131)],

$$\psi_{st} = \varepsilon\zeta\Delta P/\eta\kappa \qquad (10.88)$$

by the conductivity:

$$I_{st} = \kappa\psi_{st} = (\varepsilon\zeta/\eta)\Delta P \qquad (10.89)$$

The linear relation between I_{st} and ΔP can be expressed in terms of a single constant,

$$I_{st} = L_{I,P}\Delta P \qquad (10.90)$$

When both an applied field and a pressure difference are present and the system is fairly close to equilibrium, so that coupling between the two types of currents is minimal, the total current is the sum of the ohmic and streaming currents,

$$I = L_{I,\psi}\Delta\psi + L_{I,P}\Delta P \qquad (10.91)$$

Note that neither ψ nor P is actually a force. For the homogeneous membrane of interest, forces are formed by dividing each of these variables by d, the membrane thickness. When using the constants L, we can assume that this distance has been incorporated into the constant. If we wish to use derived expressions such as Eq. (10.89), the parameter d is included twice,

$$I_{st} = (d\varepsilon\zeta/\eta)(\Delta P/d) \qquad (10.92)$$

Equation (10.89) is preferred because d need not be known explicitly.

A flow of water through a membrane can be produced when a potential is applied across the membrane. This was the phenomenon of electroosmosis introduced in Chapter 8. This water flow is observed as an additional flow, which supplements the flow that would be produced by a pressure difference alone. Equation (8.106) defined the osmotic mobility

$$u' = \varepsilon\zeta/\eta \qquad (10.93)$$

as a velocity per unit applied potential. The resultant volume flux is then

$$J_v = u'\Delta\psi \tag{10.94}$$

or

$$J_v = (\varepsilon\zeta/\eta)\Delta\psi \tag{10.95}$$

For the applied field across the membrane of thickness d, this becomes

$$J_v = (d\varepsilon\zeta/\eta)(\Delta\psi/d) \tag{10.96}$$

The total volume flux is the sum of the electroosmotic flux and the pressure-generated flux,

$$J_v = (\varepsilon\zeta/\eta)\Delta\psi + L_{v,P}\Delta P \tag{10.97}$$

$$= L_{v,\psi}\Delta\psi + L_{v,P}\Delta P \tag{10.98}$$

The total charge flux is

$$I = L_{I,\psi}\Delta\psi + (\varepsilon\zeta/\eta)\Delta P \tag{10.99}$$

$$= L_{I,\psi}\Delta\psi + L_{I,P}\Delta P \tag{10.100}$$

The cross coefficients in these equations are obviously identical

$$L_{v,\psi} = (\varepsilon\zeta/\eta) = L_{I,P} \tag{10.101}$$

This specific identity is one example of a general set of relations known as the Onsager reciprocal relations. For a general pair of force–flux relations of the form

$$J_1 = L_{11}X_1 + L_{12}X_2 \tag{10.102}$$

$$J_2 = L_{21}X_1 + L_{22}X_2 \tag{10.103}$$

the Onsager reciprocal relation is

$$L_{12} = L_{21} \tag{10.104}$$

In the linear region under study, a net flux might be produced by the summed contributions of N different forces,

$$J_i = \sum_i L_{ij}X_j \tag{10.105}$$

Then N different fluxes are possible, because each force has a conjugate flux. There are then

$$N(N-1)/2 \tag{10.106}$$

reciprocal relations of the form

$$L_{ij} = L_{ji} \qquad i \neq j \tag{10.107}$$

The N forces and fluxes can be expressed as a vector–matrix equation,

$$\mathbf{J} = \mathbf{LX} \qquad (10.108)$$

where $\mathbf{J}$ and $\mathbf{X}$ are vectors and $\mathbf{L}$ is a matrix with elements L_{ij} for the ith row and jth column. The Onsager reciprocal relations produce a symmetric matrix.

6. OSMOTIC FLOW

The thermodynamic descriptions for osmotic pressure and the Donnan equilibrium limit the permeability of the membranes that separate the two solution phases. Osmotic pressure is calculated for a membrane that is permeable only to water. The Donnan equilibrium is determined for ion-permeable, water-impermeable membranes. We have not considered the situation where the membrane is permeable to both solvent and solute, because this requires a more involved thermodynamic analysis. However, situations where both solvent and solute are permeable are expected and are easily described using the methods of irreversible thermodynamics.

Consider the osmotic pressure generated by a water-permeable membrane in Fig. 10.1(a). A nonionic solute is present in both baths at concentrations of c_1 and c_2 respectively. If $c_2 > c_1$, water will flow to Side 2, and a hydrostatic pressure will develop on Side 2 as shown. The system reaches equilibrium when this hydrostatic pressure balances the effects of the solute concentration gradient. In our original thermodynamic discussion, we determined a single osmotic pressure. Now, we must note that there are actually two opposing pressures: a hydrostatic pressure P generated by the extra water on Side 2, and the osmotic pressure Π,

$$\Pi = (c_2 - c_1)RT \qquad (10.109)$$

This pressure will be directed toward the more concentrated solution and will be opposite to the hydrostatic pressure, as shown in Fig. 10.(b). At equilibrium, these pressures must be equal and opposite so that

$$P - \Pi = 0 \qquad (10.110)$$

Of course, this raises an interesting point. If both the hydrostatic pressure and the osmotic pressure were directed from Solution 1 to Solution 2, the resultant flow of water through the membrane would be produced by both these forces. In other words, the next flux is the sum of two forces: the

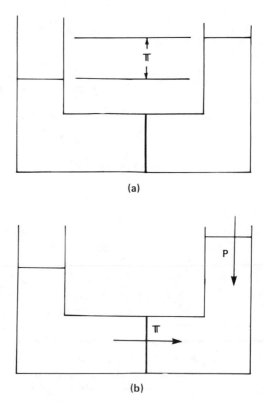

(a)

(b)

Fig. 10.1 (a) Osmotic pressure at equilibrium. The hydrostatic pressure Π, balances the concentration chemical potential. (b) Dynamic interpretation: the concentration gradient generates a driving pressure Π, which must be counterbalanced by a hydrostatic pressure P.

hydrostatic pressure is the conjugate force, while the osmotic pressure based on the solute concentration difference is the coupled force. The flux equation is

$$J_v = L_{vP}P + L_{v\Pi}\Pi \tag{10.111}$$

Since the membrane again has constant thickness d, this constant can be incorporated into the L terms and the forces are then described by the pressure differences across the membrane. Equation (10.111) is now more general than the original equilibrium expression; the pressures may contribute to a net flow when both are directed in the same direction. A net flux is possible if two opposing pressures have different magnitudes. Of course, the entire problem is complicated by the presence of the L_{ij} coefficients. If L_{vP} and $L_{v\Pi}$ have different values because of the particular nature of the membrane used, these coefficients must be included in the equilibrium expression. However, an ideal semipermeable membrane must balance both water pressures for equilibrium. The equation,

$$J_v = 0 = L_{vP}P + L_{v\Pi}\Pi \tag{10.112}$$

is consistent with this requirement only if

$$L_{vP} = L_{v\Pi} \tag{10.113}$$

That is, the water must pass through the membrane via the same pathways for either a hydrostatic or an osmotic pressure.

Two conjugate forces have been used to produce the single volume flux through the membrane. This suggests that we need a second flux that is conjugate to the osmotic pressure. The only other material that can traverse the membrane is the solute. To generate this second flux, we require a membrane that permits the flow of both water and solute.

If the membrane contains channels that permit a transmembrane solute flow, the magnitude of this flow will depend on the concentration difference between the solutions. This was derived in Chapter 9 as the solution for Fick's first law for a homogeneous membrane,

$$J = P(c_2 - c_1) = (D/d)(c_2 - c_1) \tag{10.114}$$

The gradient

$$(c_2 - c_1)/d \tag{10.115}$$

is immediately apparent. For consistency with our previous conjugate force selection, this equation must be multiplied and divided by RT to give

$$J_s = (1/RT)(D/d)[RT(c_2 - c_1)] \tag{10.116}$$

$$= (D/dRT)\Pi \tag{10.117}$$

The solute flux is proportional to the osmotic pressure.

Because the water flow has been coupled to a solute gradient, a hydrostatic pressure must produce a flow of solute for completeness. In other words, we will observe a flow of solute relative to water even when there is no solute concentration difference ($c_1 = c_2$).

To understand the nature of this coupling, consider a membrane that has separate channels for both the solute and the water. An applied hydrostatic pressure will serve to bring new solution to the membrane interfacial area. Because solute and water molecules may permeate their respective channels at different rates, we will observe an increase or decrease in the solute concentration that appears on the opposite side of the membrane. This is expressed via the coupling coefficient L_{sP} so that the solute flux equation is

$$J_s = L_{sP}P + L_{s\Pi}\Pi \tag{10.118}$$

The two flux equations are

$$J_v = L_{vP}P + L_{v\Pi}\Pi \tag{10.119}$$

$$J_s = L_{sP}P + L_{s\Pi}\Pi \tag{10.120}$$

At equilibrium, the full hydrostatic pressure occurs when the membrane is solute-impermeable. If solute can pass through the membrane with the water, the resultant hydrostatic pressure at equilibrium must be smaller than the osmotic pressure,

$$P < \Pi \tag{10.121}$$

At equilibrium, the net flux of water must be zero:

$$J_v = 0 = L_{vP}P + L_{v\Pi}\Pi \tag{10.122}$$

This equation can now be used to define P as a function of Π,

$$P = -(L_{v\Pi}/L_{vP})\Pi \tag{10.123}$$

For the ideal semipermeable membrane,

$$P = \Pi \tag{10.124}$$

at equilibrium so that

$$-(L_{v\Pi}/L_{vP}) = 1 = \sigma \tag{10.125}$$

where σ is called the reflection coefficient. For the ideal semipermeable membrane it attains its maximum value of 1. Physically, $\sigma = 1$ signifies the reflection of all solute molecules from the membrane.

When $\sigma < 1$, some solute permeation is also possible and the amount of permeation increases as σ approaches zero. The special case $\sigma = 0$ is significant. Equation (10.125) defines σ for all cases. If the reflection coefficient is zero, $L_{v\Pi} = 0$. For this case, water and solute are equally permeable. The membrane is fully permeable and no hydrostatic pressure can develop from the solute concentration gradient.

The reflection coefficient can become negative for cases in which the solute permeability pathways permit the solute to diffuse more rapidly than the water. The negative sign creates a situation in which both P and Π have the same, rather than opposing, signs. The solute flows rapidly from its high concentration to its lower concentration, and water follows instead of flowing to the region of higher solute concentration.

7. SAXEN'S RELATIONSHIPS

For two conjugate forces and fluxes, the linear equations can be written as

$$J_1 = L_{11}X_1 + L_{12}X_2 \qquad (10.126)$$

$$J_2 = L_{21}X_1 + L_{22}X_2 \qquad (10.127)$$

The Onsager relation

$$L_{12} = L_{21} \qquad (10.128)$$

holds for these equations. This relation permits us to establish a variety of relations between the forces and the fluxes of Eqs. (10.126) and (10.127).

If the force X_2 is eliminated—that is, $X_2 = 0$—from both equations, they become

$$(J_1)_{X_2=0} = L_{11}X_1 \qquad (10.129)$$

$$(J_2)_{X_2=0} = L_{21}X_1 \qquad (10.130)$$

and the ratio of the two expressions is

$$(J_2/J_1) = L_{21}/L_{11} \qquad \text{for} \quad X_2 = 0 \qquad (10.131)$$

Because $L_{21} = L_{12}$, the ratio of fluxes will also be equal to any ratio of the form

$$L_{12}/L_{11} \qquad (10.132)$$

This ratio can be equated to a ratio of forces by setting $J_1 = 0$ in Eq. (10.126):

$$0 = L_{11}X_1 + L_{12}X_2 \qquad (10.133)$$

or

$$L_{12}/L_{11} = -(X_1/X_2)_{J_1=0} \qquad (10.134)$$

The ratios of forces and fluxes can now be equated to give

$$(J_2/J_1)_{X_2=0} = -(X_1/X_2)_{J_1=0} \qquad (10.135)$$

which is one of Saxen's relationships.

The ratios of fluxes and forces are equal under certain conditions, that is, when $X_2 = 0$ and $J_1 = 0$.

The two equations can be manipulated to get a second relationship,

$$(J_1/J_2)_{X_1=0} = -(X_2/X_1)_{J_2=0} \qquad (10.136)$$

An additional relationship can be developed by solving the individual flux equations for L_{12} and L_{21}, respectively:

$$L_{12} = (J_1/X_2)_{X_1=0} \tag{10.137}$$

and

$$L_{21} = (J_2/X_1)_{X_2=0} \tag{10.138}$$

Because the two coefficients are equal, we find

$$(J_1/X_2)_{X_1=0} = (J_2/X_1)_{X_2=0} \tag{10.139}$$

another Saxen relationship.

The form of this equation parallels that for the Maxwell equations in reversible thermodynamics. For example,

$$(\partial V/\partial T)_P = -(\partial S/\partial P)_T \tag{10.140}$$

Note the interchange of the variables of differentiation in both cases.

Equation (10.139) suggests a potential extension of Eq. (10.139) to

$$(\partial J_1/\partial X_2)_{X_1=0} = (\partial J_2/\partial X_1)_{X_2=0} \tag{10.141}$$

In the region where the linear flux–force relations are valid, the two derivatives would reduce to L_{12} and L_{21}, respectively. However, Eq. (10.141) might remain valid in regions further removed from equilibrium where the linear flux–force relations are no longer valid. Equation (10.141) has not been proven by either theoretical or experimental methods.

To illustrate the utility of these methods, we consider the special case of a gramacidin-A channel in a bilayer membrane. Gramicidin-A dimerizes to form a channel that permits both water and cation flow. Rosenberg and Finkelstein (1978) used the streaming potential produced by an osmotic pressure gradient to calculate the ratio of water to ion in the channel. If we use the osmotic pressure Π rather than the hydrostatic pressure P, the equations for volume flux and charge flux are

$$J_v = L_{v\Pi}\Pi + L_{v\psi}\psi \tag{10.142}$$

$$I = L_{I\Pi}\Pi + L_{I\psi}\psi \tag{10.143}$$

Note that the $\Delta\psi$ term, the streaming potential, has been replaced by a potential difference ψ. Both forces in the equations are still differences between the two baths. Equation (10.139) becomes

$$(J_v/I)_{\Pi=0} = (\psi/\Pi)_{I=0} \tag{10.144}$$

or

$$\psi = (J_v/I)_{\Pi=0}\Pi \tag{10.145}$$

The streaming potential and the osmotic pressure are experimentally measurable, and permit determination of the ratio J_v/I. However, this ratio must now be converted into a ratio of moles of water to moles of ion as both pass through the gramicidin channel.

The molar volume of water $\overline{V}$ is the volume for 1 mole of water. Dividing J_v by this molar volume gives the moles of water flowing through the channel,

$$n_w = J_v/\overline{V} = \text{moles } H_2O/\text{area} \cdot s \qquad (10.146)$$

The variable I describes the coulombs of charge per unit area per second. To convert this to units of moles of ion per second per area, we divide by the Faraday constant,

$$n_K = I/F \quad \text{(moles of ion/area} \cdot s) \qquad (10.147)$$

We define N' as the ratio of these two fluxes,

$$N' = n_w/n_K = (J_v F/I\overline{V}) \qquad (10.148)$$

Equation (10.145) can now be written in terms of N' by introducing $\overline{V}$ and F,

$$\psi = (FJ_v/I\overline{V})(\overline{V}\Pi/F) = N'(\overline{V}\Pi/F) \qquad (10.149)$$

Because $\overline{V}$ and F are known, a measurement of the streaming potential for a given osmotic pressure difference will give N', the ratio of water molecules to ions passing through the gramicidin channel. Rosenberg and Finkelstein estimated a ratio of 6.5 water molecules to each ion in the channel using this technique.

8. THE CURIE SYMMETRY PRINCIPLE

For the relations developed using the Onsager reciprocal relations, we have always selected vector forces and fluxes. Scalar forces and fluxes such as those describing chemical reactions may also be present in the system. We might wish to extend our force–flux relations to expressions of the form

$$J_3 = L_{11}X_1 + L_{12}A \qquad (10.150)$$

$$\xi = L_{21}X_1 + L_{22}A \qquad (10.151)$$

where X_1 is a vector force.

This set of coupled equations poses an interesting problem. If a reaction does take place in the system, then Eq. (10.150) suggests that this reaction will lead to a net flux in or out of the system. However, chemical reaction takes place in no specific direction. If an excess of material were generated within the system, it would appear at equal rates everywhere within the system and no concentration gradients would develop within this homogeneous system. No directional flows of material are possible within the system. For this reason, the coupling coefficients L_{12} and L_{21} in Eqs. (10.150) and (10.151) must be zero. This lack of coupling is called the Curie symmetry principle: tensors of different rank will not couple in a homogeneous system. The scalar fluxes and forces of the chemical reaction are tensors of zero order, while vector forces and fluxes are first-order tensors.

If the chemical reaction proceeds homogeneously in the system, some chemical species must be augmented at the expense of other species. If the reaction produces a concentration of some species that is larger than the external concentration of this species, a concentration gradient can develop across the system boundary. Thus, a vector flow from the interior to the exterior is possible if the system boundary, such as a membrane, permits such a flow. The chemical reaction appears to give a vector flux in violation of the Curie symmetry principle.

The Curie symmetry principle is valid only for homogeneous systems. The boundary separates two distinct phases and therefore creates an inhomogeneity. The membrane must have some special properties to permit a coupling between the two regions. For example, if the membrane were equally permeable to reactants and products, these species would establish a homogeneous arrangement on both sides of the membrane. Because the membrane exerts an equal influence on all species, it can be removed leaving one continuous homogeneous region.

The Curie symmetry principle is consistent with our general experience for chemical reactions. For the reaction

$$A \longrightarrow B \qquad (10.152)$$

we do not observe a spatial separation in which A moves left while B moves right in the system. However, there are chemical reactions that produce such separations in homogeneous media. The most well known of these reactions is the Belousov–Zhabotinsky reaction, which is actually a combination of a number of reaction steps. The reaction is initiated by mixing malonic acid with cerium sulfate and potassium bromate. When sufficient bromate is present, the complicated mechanism of the reaction can be summarized with the two reactions

$$CH_2(COOH)_2 + 6\,Ce^{4+} + 2\,H_2O \longrightarrow 2\,CO_2 + HCOOH + 6\,Ce^{3+} + 6\,H^+ \qquad (10.153)$$

and

$$4\,Ce^{3+} + BrO_3^- + CH_2(COOH)_2 + 5\,H^+ \longrightarrow BrCH(COOH)_2 + 4\,Ce^{4+} + 3\,H_2O \qquad (10.154)$$

In the first reaction, Ce^{4+} is reduced to Ce^{3+}, while Ce^{3+} is oxidized to Ce^{4+} in the second reaction. The reaction system exhibits oscillating concentrations of Ce^{4+} and Ce^{3+}. These oscillations are energized via the monotonic decrease of both malonic acid and bromate ion concentrations with time.

In addition to this oscillatory kinetic behavior, the Belousov–Zhabotinsky reaction also generates spatial organization; the concentrations of species do not remain homogeneous during the reaction process. The reaction produces patterns generated by the spatially distinct Ce^{3+} and Ce^{4+} regions within the system, as shown in Fig. 10.2. This particular set of reactants then produces concentration gradients within the homogeneous medium.

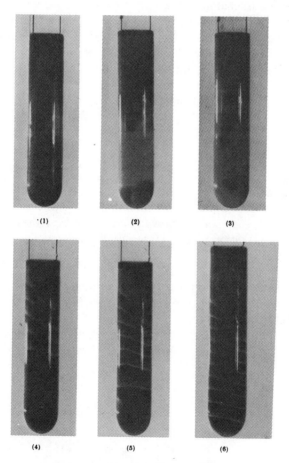

Fig. 10.2 Establishment of the dissipative structure. [Reproduced with kind permission of Prigogine, 1971.]

9. STEADY-STATE COUPLING

If a membrane with some limiting property separates two phases, the Curie symmetry principle will break down, and it becomes possible to couple the scalar reactions process with other vector processes. The membrane then acts to produce a net flow so that the scalar reaction process within the system is converted to a net flux out of the system.

To illustrate this process, we consider the special case of a membrane that is permeable to both the reactant R and the product P of a reaction. The membrane is diathermal so that heat cannot pass through the membrane. This permits us to maintain the two phases at different temperatures— that is, we have created an inhomogeneity by limiting heat flow. The heat cannot flow, but it can induce a flow of chemical species across the boundary.

The situation is described in Fig. 10.3. The membrane separates two phases with temperatures T_1 and T_2, respectively. Because the rate constants for reaction will differ with temperature, the rates of reaction on both sides will differ. The first-order rate constants for the reactions

$$R_1 \underset{m_1}{\overset{k_1}{\rightleftharpoons}} P_1 \tag{10.155}$$

and

$$R_2 \underset{m_2}{\overset{k_2}{\rightleftharpoons}} P_2 \tag{10.156}$$

are k_1 and m_1 and k_2 and m_2, respectively. If k_2 is larger than k_1, the concentration of P_2 will rise more rapidly than that of P_1. Because the membrane is permeable to P, species P will flow to Side 1. However, the ratio of $[P]/[R]$ must remain constant on Side 1 via the equilibrium relation

$$K(T_1) = k_1/m_1 = [P_1]/[R_1] \tag{10.157}$$

As $[P_1]$ increases due to flow across the membrane, P will be converted to R on Side 1. From the three processes considered, R on Side 2 has been transferred to Side 1. A concentration gradient for R now exists but is opposite to that observed for P. Because the membrane is also permeable to R, R will flow from Side 1 to Side 2 to complete the fourth leg of a cycle. A particle,

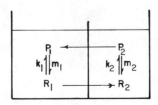

Fig. 10.3 Steady-state coupling for baths at different temperatures.

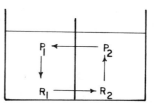

Fig. 10.4 The generation of a steady cyclic flow of particles in a system with a temperature gradient.

starting as reactant on Side 2 eventually reappears as reactant on Side 2 after passing through the intermediate states P_2, P_1, and R_1.

The cyclic flow will persist as long as the temperature gradient between the phases is maintained. The concentrations of R_1, R_2, P_1, and P_2 will remain constant, but these constant values will be dictated by the reaction rates and permeabilities of the species. The tools to calculate these concentrations and flow rates are introduced in Chapter 11.

Because the concentrations are dictated by the temperature gradient, this cycling system is not at equilibrium even though the concentrations of each species remain constant. The model describes a steady-state situation in which the flow of particles through each state remains constant, as shown in Fig. 10.4. For example, the number of P_2 molecules crossing the membrane to form P_1 molecules is exactly equal to the number of P_1 molecules which react to form R_1 molecules. The coupling between the temperature gradient and the reaction processes is called steady-state coupling. Steady-state coupling utilizes some selective property of the membrane to induce a cyclic flow.

Chemical reactions may also be coupled to produce a net flow between two phases. Consider the reaction

$$R \longrightarrow P$$

again. If both species can traverse the membrane, their fluxes through the membrane will be dictated by their chemical potential difference between both sides of the membrane,

$$\Delta\mu_R = \mu_R(2) - \mu_R(1) \tag{10.158}$$

and

$$\Delta\mu_P = \mu_P(2) - \mu_P(1) \tag{10.159}$$

An additional, nonreacting molecule I is also present with a chemical potential difference $\Delta\mu_I$. The membrane is permeable to all three species. In order to generate inhomogeneity, we assume that I can couple with P but not with R. A flow of P from the system will induce a concomitant flow

of I. By symmetry, the flux of P will depend on the flux of I. The flux equations
for the three species are

$$J_P = L_{PP}\Delta_P + L_{PI}\Delta\mu_I \qquad (10.160)$$

$$J_I = L_{IP}\Delta\mu_P + L_{II}\Delta\mu_I \qquad (10.161)$$

$$J_R = L_{RR}\Delta\mu_R \qquad (10.162)$$

In the absence of reaction between R and P, these three fluxes would approach
zero—that is, equilibrium between the phases would result. If reaction is
included, a steady flow from R to P is generated. The increased concentration
of P produces an increased flux of P from the system via an increased chemical
potential. Because of the coupling, an increased flux of I is expected. However,
we must establish the direction of this flux.

At a steady state, the flux from R to P due to reaction must be equal to
the flux J_P so that the concentration of P in the system remains constant.
To keep the concentration of R constant, R must enter the system at the same
rate. All three fluxes are equal, as shown in Fig. 10.5. The molecule I under-
goes no reaction. When the chemical potential of P changes because of the
reaction, the chemical potential of I must adjust accordingly so that the net
flux of I, J_I, remains zero. Because of the I–P coupling, the flux of I depends
on P and the balance of the two chemical potentials hold the flux at zero in
the steady state.

The flux equations are now

$$J = L_{PP}\Delta\mu_P + L_{PI}\Delta\mu_I \qquad (10.163)$$

and

$$0 = L_{IP}\Delta\mu_P + L_I\Delta\mu_I \qquad (10.164)$$

The chemical potentials are determined by solving these two simultaneous
equations. The equations are solved using the method of determinants.
For $\Delta\mu_I$,

$$\Delta\mu_I = \begin{vmatrix} L_{PP} & J \\ L_{IP} & 0 \end{vmatrix} \bigg/ \begin{vmatrix} L_{PP} & L_{PI} \\ L_{IP} & L_{II} \end{vmatrix} \qquad (10.165)$$

$$= (-L_{IP}J)/(L_{PP}L_{II} - L_{IP}L_{PI}) \qquad (10.166)$$

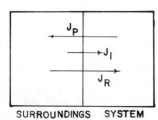

Fig. 10.5 Steady-state coupling involving reactive species R and P and a coupled
nonreactive species I.

SURROUNDINGS   SYSTEM

The chemical potential is directed against the cyclic flux that is, directed into the system. The reaction has produced an increased P flux out of the system due to an increased chemical potential in this direction. The chemical potential of I acts as a counterbalance to keep the net flux of I equal to zero.

The chemical potential for P is

$$\Delta\mu_P = \begin{vmatrix} J & L_{PI} \\ 0 & L_{II} \end{vmatrix} \Big/ (L_{PP}L_{II} - L_{IP}L_{PI})$$

$$= +JL_{II}/(L_{PP}L_{II} - L_{IP}^2) \tag{10.167}$$

using the Onsager reciprocal relation,

$$L_{IP} = L_{PI} \tag{10.168}$$

The chemical potential for P points in the same direction as the net flux, as expected.

Equation (10.167) describes a direct proportionality between the force $\Delta\mu_P$ and the flux J_P. If a single system containing R, P, and I were examined, the observer would find a direct proportionality

$$J' = L\Delta\mu_P \tag{10.169}$$

and might conclude there was no coupling. To resolve this problem, the experiment would have to be repeated without I. In this case,

$$J_P = L_{PP}\Delta\mu_P \tag{10.170}$$

and

$$L_{PP} \neq L = (L_{PP}L_{II} - L_{IP}^2)/L_{II} \tag{10.171}$$

with L determined from Eq. (10.167). The two coefficients do become equal in the limit of no coupling, that is,

$$L_{IP} \to 0 \tag{10.172}$$

PROBLEMS

1. For the pair of flux equations,

$$J_v = L_{11}\Delta P + L_{12}\Delta\psi$$
$$J_q = L_{21}\Delta P + L_{22}\Delta\psi$$

derive Saxen's relations.

2.   In a membrane system, a concentration gradient and a pressure gradient are maintained across the membrane. (a) Calculate the form of the flux equations for this system. (b) Determine the dissipation function. (c) Explain the physical significance of the cross coupling for this system.

3.   When an ion traverses a channel, it carries n water molecules with it. Use irreversible thermodynamics to estimate n in terms of the coupling coefficients, L. The waters constitute a volume flow, and the ions constitute a current.

4.   Discuss the difference between a steady state and equilibrium. In this context, discuss the notion that all systems must ultimately come to equilibrium.

5.   Linear nonequilibrium thermodynamics relates the entropy dissipation function $T\,dS/dt$ to a sum of products of flows and forces. For example,

$$T\,dS/dt = J_{s}[(d/dx)(T)] + J_{m}[(d/dx)(-\mu)]$$

where J_s and J_m are the entropy and mass fluxes. Also,

$$J_i = \sum L_{ij} X_j$$

Use this information to establish a relationship between the L terms and the diffusion coefficient.

Chapter 11

Kinetics

1. BASIC KINETICS

In previous chapters, we have examined transport across membranes using diffusion equations and the equations of irreversible thermodynamics. In most of these studies, we observe a steady flow of material through the membrane. However, such flows are often controlled by chemical processes within the membrane or channel. For example, a permeant ion may move to a binding site within the channel at some rate; it may also dissociate or be displaced from this site at some rate. These rates of binding and dissociation then determine the time required for the ion to traverse the channel.

A simple kinetic process involves a large number of independently reacting molecules. The reaction of one molecule has no effect on the time of reaction for other molecules. At any instant, the rate of reaction is directly proportional to the number of molecules available to react. This can be written as a first-order rate equation,

$$dc/dt = -kc \tag{11.1}$$

where k, the rate constant, determines the rate at which these independent molecules react. Because the equation is first order, we must specify one boundary condition. In this case, we specify the initial concentration

$$c(0) = c_o \tag{11.2}$$

The equation is solved by dividing both sides by c and integrating,

$$\int (dc/c) = -\int k \, dt \tag{11.3}$$

$$\ln c = -kt + C \tag{11.4}$$

267

When the initial condition is substituted into this equation, it becomes

$$\ln c_0 = -k(0) + C \tag{11.5}$$

so that

$$\ln[c(t)] = -kt + \ln c_0 \tag{11.6}$$

Collecting logarithmic terms and exponentiating both sides gives

$$c(t) = c_0 \exp(-kt) \tag{11.7}$$

When each molecule is free to react independently, the molecular concentration will decay exponentially.

The term "independent" must be used quite carefully in this context. The molecule probably acquired the energy for its transformation from neighboring molecules, such as water molecules in solution. Because there are such a large number of these molecules available and because they do not change their molecular state, the details of their role in the reaction is averaged into the rate constant. This constant is a quantitative measure of the time it took the molecule to acquire the energy to react.

Although this rate constant is sufficient to describe the kinetics of this first-order system, an alternative statement in terms of time is often used. The lifetime τ is defined as the time required for the molecular concentration to reach a concentration of c_0/e where e is the exponential. When this value is substituted into Eq. (11.7), we find

$$[(c_0/e)/c_0] = e^{-1} = \exp(-k\tau) \tag{11.8}$$

This equation will be satisfied when

$$k\tau = 1 \tag{11.9}$$

so

$$\tau = 1/k \tag{11.10}$$

Because c_0 cancels completely on the left-hand side of Eq. (11.8), a first-order equation will always be independent of the starting concentration. In fact, this is one of the simplest ways of determining whether a reaction is first order. When the lifetime is constant for arbitrary initial concentrations, the reaction will be first order.

The first-order reaction presented here implies that the molecules react irreversibly. However, if a molecule A reacts to form molecule B, detailed balance and microscopic reversibility always require a reverse reaction as well. In this case, the differential equation (with A as number of molecules of A and B as number of molecules of B) expands to a pair of differential equations,

$$dA/dt = -k_f A + k_r B \qquad dB/dt = k_f A - k_r B \tag{11.11}$$

Although solutions for coupled equations of this type will be developed in later sections, this particular system can be reduced to a single first-order equation by noting that conservation of mass requires a constant sum of A and B molecules, that is,

$$M = A(t) + B(t) \tag{11.12}$$

at any time during the reaction. This relation can be used to eliminate B from the first differential equation,

$$dA/dt = -k_f A + k_r(M - A) = k_r M - (k_f + k_r)A \tag{11.13}$$

The resultant integrals have the form

$$\int \{dA/[k_r M - (k_f + k_r)A]\} = \int dt \tag{11.14}$$

This integral is identical to the one that appeared for the Goldman equation in Chapter 9. The equation integrates as

$$\frac{\ln[k_r M - (k_f + k_r)A]}{-(k_f + k_r)} = t + \text{constant} \tag{11.15}$$

Assume that there are no B molecules at time zero, that is, $M = A_o$. The constant can now be evaluated as

$$\ln[k_r A_o - (k_f + k_r)A_o] = -(k_f + k_r)0 + C' = \ln(-k_f A_o) = C' \tag{11.16}$$

When this constant is substituted into the equation and terms containing A and A_o are collected on the left before exponentiating both sides, we find

$$- [k_r A_o - (k_f + k_r)A]/k_f A_o = \exp[-(k_f + k_r)t] \tag{11.17}$$

Note that A_o is substituted for M, because the total number of molecules will always be A_o, the number with which the reaction started. This equation can be rearranged to

$$A(t) = \{[k_r/(k_f + k_r)] + [k_f/(k_f + k_r)] \exp[-(k_f + k_r)t]\}A_o \tag{11.18}$$

For long times, this concentration reduces to the equilibrium concentration of A,

$$A(t) = [k_r/(k_f + k_r)]A_o \tag{11.19}$$

The fraction of the total concentration for A molecules at equilibrium is directly proportional to the rate constant that leads into that configuration.
The concentration $B(t)$ can be determined, using the relation

$$B(t) = A_o - A(t), \text{ as}$$

$$B(t) = [k_f/(k_f + k_r)]\{1 - \exp[-(k_f + k_r)t]\}A_o \tag{11.20}$$

When more than two molecules are explicitly involved in a given reaction, the reaction becomes bimolecular, and this generally leads to a second-order differential equation. To illustrate the difference between first- and second-order reactions, we consider a simple example in which molecule A reacts with a second A molecule to form some product. The reverse reaction is ignored. If the initial concentration of A is A_o, the differential equation will be

$$dA/dt = -kA^2 \qquad A(0) = A_o \qquad (11.21)$$

The rate constant k will have the units of (concentration)$^{-1}$s^{-1}, for example, (moles/liter)$^{-1}$s^{-1}. The equation can be arranged to the form

$$dA/A^2 = -k\,dt \qquad (11.22)$$

and integrated to give

$$-1/A = -kt + C \qquad (11.23)$$

The constant is evaluated from the initial conditions as

$$-1/A_o = C \qquad (11.24)$$

and the final result is

$$(1/A) - (1/A_o) = kt \qquad (11.25)$$

In this case, the lifetime will be dependent on the initial concentration,

$$\tau = (1 - e)/kA_o \qquad (11.26)$$

Because a first-order reaction is the only reaction that does not depend on the initial concentration, this information alone does not determine the reaction order. If $1/A$ is plotted versus time and the resultant plot is a straight line, the second-order reaction can be distinguished from other possible reaction orders.

When the second-order reaction involves two different species that have different initial concentrations, the problem becomes more complicated again. Consider the reaction of A and B with initial concentrations A_o and B_o. The preferred way to solve such a kinetic equation is to define a variable x as the concentration of each species consumed. Each time a pair of molecules is consumed, the concentration x will increase, so the differential equation is written as

$$dx/dt = k(A_o - x)(B_o - x) \qquad (11.27)$$

The differential equation has the form

$$dx/[(A_o - x)(B_o - x)] = k\,dt \qquad (11.28)$$

The left-hand side can be integrated by converting it into a combination of the form

$$[C\,dx/(A_o - x)] + [D\,dx/(B_o - x)] = k\,dt \qquad (11.29)$$

Either of these terms is easily integrated once the constants C and D are determined.

These constants can be easily evaluated when the terms in the denominator are raised to the first power. The constant C can be evaluated by multiplying the factor

$$1/[(A_o - x)(B_o - x)] \qquad (11.30)$$

by $(A_o - x)$, the factor that will appear in the denominator, and then by evaluating the function that remains at $x = A_o$:

$$C = (A_o - x)[1/(A_o - x)(B_o - x)]_{x=A_o} = [1/(B_o - A_o)] \quad (11.31)$$

The constant D is evaluated similarly as

$$D = [(B_o - x)\{1/(A_o - x)(B_o - x)\}]|_{x=B_o} = 1/(A_o - B_o) \quad (11.32)$$

Equation (11.29) can now be integrated. The initial condition will be $x(0) = 0$. The integrated result is

$$-C\ln(A_o - x) - D\ln(B_o - x) = kt + C' \qquad (11.33)$$

where C' is determined from the initial condition

$$x(0) = 0 \qquad (11.34)$$

With equations for C and D from Eqs. (11.31) and (11.32),

$$C' = (B_o - A_o)^{-1}[\ln(B_o - 0) - \ln(A_o - 0)] \qquad (11.35)$$

$$= (B_o - A_o)^{-1}[\ln(B_o/A_o)] \qquad (11.36)$$

The final solution of the kinetic equation is then

$$(B_o - A_o)^{-1}\{\ln[(B_o - x)/(A_o - x)]\} - \ln(B_o/A_o) = kt \qquad (11.37)$$

$$(B_o - A_o)^{-1}\{\ln[A_o(B_o - x)/B_o(A_o - x)]\} = kt \qquad (11.38)$$

Even though reverse reaction has not been included for this kinetic example, the result is quite complicated. For this reason, it is often convenient to use alternate kinetic techniques to study such systems.

2. RELAXATION KINETICS

Consider a reaction

$$A + B \xrightleftharpoons[m]{k} C \qquad (11.39)$$

for which both forward and reverse rate processes are possible. The forward and reverse rate constants are k and m, respectively. Because the irreversible reaction of this type has an involved solution, this reversible case is obviously much more complex. However, when this reversible reaction reaches equilibrium, its rate must be zero:

$$(dC/dt)_{eq} = 0 = k[A][B] - m[C] \qquad (11.40)$$

The ratio of concentrations at equilibrium is then

$$\{[C]/[B][A]\}_{eq} = k/m \qquad (11.41)$$

Thus, the equilibrium constant is related to the ratio of the forward and reverse rate constants,

$$K_{eq} = k/m \qquad (11.42)$$

This the law of Waage and Guldberg.

Because the rate approaches zero as the system approaches equilibrium, we can describe systems near equilibrium in terms of a linear deviation from equilibrium. This is the basis of a variety of relaxation methods. The system is perturbed from equilibrium, and its temporal return to equilibrium is then observed. By perturbing the system to a state near equilibrium, we can produce a kinetic equation of first order in the concentration.

To illustrate the technique, consider an arbitrary perturbation of Eq. (11.39) that increases the concentration of C by an amount x,

$$[C] = [C]_{eq} + x \qquad (11.43)$$

The concentrations of both A and B must then fall by x,

$$[A] = [A]_{eq} - x \qquad (11.44)$$

$$[B] = [B]_{eq} - x \qquad (11.45)$$

These perturbed values are now substituted into the rate equation [Eq. (11.40)],

$$d\{[C]_{eq} + x\}/dt = k\{[A]_{eq} - x\}\{[B]_{eq} - x\} - m\{[C]_{eq} + x\} \qquad (11.46)$$

Collecting terms in powers of x, we find

$$d[C]_{eq}/dt + dx/dt = k[A]_{eq}[B]_{eq} - m[C]_{eq} - k\{[A]_{eq} + [B]_{eq}\}x$$
$$-mx + kx^2 \qquad (11.47)$$

At equilibrium,

$$d[C]_{eq}/dt = k[A]_{eq}[B]_{eq} - m[C]_{eq} = 0 \qquad (11.48)$$

and these terms drop from each side of the equation. If the perturbation from equilibrium is small,

$$x^2 \ll x \qquad (11.49)$$

and the square term can be ignored. The resultant kinetic equation is then first order in x, the deviation from equilibrium:

$$dx/dt = -(k\{[A]_{eq} + [B]_{eq}\} + m)x \qquad (11.50)$$

This first-order equation will have a solution of the form

$$x = x_o \exp(-k't) \qquad (11.51)$$

where x_o is the magnitude of the fluctuation from equilibrium at $t = 0$ and

$$k' = k\{[A]_{eq} + [B]_{eq}\} + m \qquad (11.52)$$

For perturbations near equilibrium, any complicated kinetic equation can be reduced to a simple first-order equation. Such perturbations will arise naturally in chemical systems with no external input. The same techniques can be used to describe these "noise" fluctuations.

3. KINETIC ACTIVATION ENERGY

In thermodynamics, the energy and free energy of a reaction describe energetic changes produced by breaking some chemical bonds and producing others. Although our linear analysis of chemical energetics in Chapter 10 indicated that the degree of advancement was directly proportional to the free energy difference, this direct proportionality was valid only when the system was near equilibrium. In general, the free energy difference has no apparent connection with the rate. For example, the free energy for the formation of water from hydrogen and oxygen is large but the rate of water formation is low, because the molecules need a surface or third body to facilitate reaction. To describe such differences, we need additional expressions for kinetic energetics.

For a reaction system at equilibrium in a closed container, we can determine the equilibrium constant at different temperatures as a function of the internal energy of the reaction using the Gibbs–Helmholtz equation,

$$\ln[K(T_2)/K(T_1)] = -(\Delta H/R)[(1/T_2) - (1/T_1)] \qquad (11.53)$$

For a single temperature,

$$\ln K = -(\Delta H/R)(1/T) \qquad (11.54)$$

where ΔH is the enthalpy for the reaction.

In Section 2 we stated the law of Waage–Guldberg, which related the forward and reverse rate constants to the equilibrium constant

$$K = k/m \qquad (11.55)$$

To determine a relationship similar to Eq. (11.54) for the forward and reverse rate constants, we postulate an energy barrier of height E_a between the reactants and the products, as shown in Fig. 11.1. The height of the barrier will now limit the rates of interconversion of reactants and products. Note that the barrier for reverse reaction is larger in Fig. 11.1, because the products lie ΔH below the reactants in the figure.

Because Eq. (11.55) provides a connection between thermodynamics and kinetics, we use it in conjunction with the Gibbs–Helmholtz equation to give

$$\ln K = \ln k - \ln m = -(\Delta H/R)(1/T) \qquad (11.56)$$

which is equivalent to

$$\ln k - \ln m = -(E_a + \Delta H - E_a)/RT$$
$$= (-E_a/RT) + (E_a - \Delta H)/RT \qquad (11.57)$$

Because E_a is obviously the barrier energy for the forward reaction, we postulate the following expressions for the temperature dependence of the rate constants:

$$\ln k = -E_a/RT \qquad (11.58)$$

$$\ln m = -(E_a - \Delta H)/RT \qquad (11.59)$$

Note that a negative ΔH will increase the barrier height in Eq. (11.59).

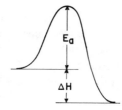

Fig. 11.1 Arrhenius activation energies E_a and $E_a + \Delta H$ for forward and reverse rate constants.

These rate constant expressions are called Arrhenius expressions and are often expressed in exponential form,

$$k = A \exp(-E_a/RT) \tag{11.60}$$

where A, the frequency factor, is the constant of integration for the system.

By analogy to the Gibbs–Helmholtz equation, the activation energy associated with a given rate constant can be determined if the rate constant is known for a series of temperatures. A plot of $\ln[k(T)]$ versus $1/T$ will give a slope

$$-E_a/R \tag{11.61}$$

For example, the kinetics for the channels in squid giant axon are described by

$$Q_{10} = 3 \tag{11.62}$$

which means that the rate constants for the processes triple when the temperature is raised $10°C$. Since the Hodgkin–Huxley rate constants were tabulated at $6.3°C$, a particular rate constant k at $6.3°C$ will become $3k$ at $16.3°C$. Thus,

$$\ln(3k/k) = \ln 3 = -(E_a/R)[(1/289.5 \text{ K}) - (1/279.5 \text{ K})] \tag{11.63}$$

and the activation energy is

$$E_a = 17.5 \text{ kcal/mol}$$

4. TRANSITION-STATE THEORY

Transition-state theory provides one technique for the determination of both the frequency factor A and the activation energy for a given reaction. This theory was introduced in Chapter 9 when we discussed the Parlin–Eyring model for ion transport through a membrane. In that discussion, we noted that the rate constants describing transitions were related to the height of a free energy barrier. In the present discussion, the activation energy was related to the internal energy of the system. We can now clarify the relationship between these two descriptions.

Transition-state theory postulates an equilibrium between the reactant(s) and some metastable activated complex or transition state. The equilibrium constant $K^{\ddagger}$ is related to $G^{\ddagger}$, the height of the free energy barrier (Fig. 11.2), as

$$G^{\ddagger} = -RT \ln K^{\ddagger} \tag{11.64}$$

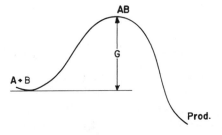

Fig. 11.2 $G^{\ddagger}$ as the free energy for activated complex formation from reactants. Reaction proceeds from the complex to products.

Although the reactant and the activated complex are assumed to satisfy thermodynamic equilibrium, the complex is metastable. Although it remains in dynamic equilibrium with the reactant, some of the metastable complexes cross the barrier and separate to form product, as shown in Fig. 11.2. The rate of "bleeding" to products determines the rate constant for the reaction.

The metastable transition state will decay in one vibrational period. The complex forms and, if it vibrates along bonds which lead to product, the complex will separate to form products. The time required for such a vibration-to-translation transition is determined by equating the thermal energy of the complex with its vibrational energy; that is, the thermal energy of the system provides the energy for the vibration of the complex,

$$h\nu = k_b T \tag{11.65}$$

where ν is the vibrational frequency and k_b is Boltzmann's constant. The frequency at which the complex decays is then

$$\nu = k_b T/h \tag{11.66}$$

This frequency is approximately 10^{13} s^{-1}, the frequency range for vibrations.

The frequency functions as a measure of the rate of product formation for a situation in which there is no energy barrier. For an activation barrier of energy $G^{\ddagger}$, the rate constant can be written for comparison with the Arrhenius equation,

$$k = (k_b T/h)[\exp(-G^{\ddagger}/RT)] \tag{11.67}$$

where the double dagger has been added to express the energy of the transition state relative to the reactants. (This dagger symbol was originally a printer's error but has now become the standard symbol for a transition state.)

If the transition-state concentration is defined as $[C^{\ddagger}]$, the equilibrium between a pair of reactants A and B with $C^{\ddagger}$ is

$$K^{\ddagger} = [C^{\ddagger}]/[A][B] \tag{11.68}$$

The concentration of activated complex is then

$$[C^\ddagger] = K^\ddagger[A][B] \tag{11.69}$$

and the rate of product formation is determined by the frequency of dissociation of this complex,

$$\text{rate} = v[C^\ddagger] = (k_b T/h)K^\ddagger[A][B] \tag{11.70}$$

The rate constant for the reaction is then the product of the frequency factor and the equilibrium constant,

$$k = (k_b T/h)K^\ddagger \tag{11.71}$$

Equation (11.64) is now used to express this equilibrium constant in terms of free energy of the barrier,

$$K^\ddagger = \exp(-G^\ddagger/RT) \tag{11.72}$$

The rate constant is

$$k = (k_b T/h)[\exp(-G^\ddagger/RT)] \tag{11.73}$$

The equilibrium constant has been expressed as a ratio of concentrations. However, it is also equivalent to the ratio of the partition functions for the molecules A and B and the activated complex. Thus,

$$K^\ddagger = Q^\ddagger/Q_A Q_B \tag{11.74}$$

where the Q_i are the canonical partition functions for each molecular species.

Although these forms of the rate constant are most common, they are only valid when the standard states have values of unity. For an arbitrary standard state, the equation must be modified to include the standard concentrations

$$G^\ddagger = -RT \ln[K^\ddagger/(c^o)^{\Delta n}] \tag{11.75}$$

where c^o is the standard-state concentration and Δn is the change in moles during formation of the activated complex. For

$$A + B \longrightarrow C^\ddagger,$$

$$\Delta n = -1 \tag{11.76}$$

For an mth order reaction,

$$\Delta n = 1 - m \tag{11.77}$$

the rate-constant expression is

$$k = (k_b T/h)[\exp(-G^\ddagger/RT)](c^o)^{1-m} \tag{11.78}$$

We can now establish a relation between transition-state theory and the Arrhenius expression. Because

$$G^{\ddagger} = H^{\ddagger} - TS^{\ddagger} \tag{11.79}$$

the equation can be separated into energetic and entropic components,

$$k = (k_b T/h)[\exp(S^{\ddagger}/R)\exp(-H^{\ddagger}/RT)](c^{\circ})^{1-m} \tag{11.80}$$

Comparing this with the Arrhenius expression, we find

$$E_a = H^{\ddagger} \tag{11.81}$$

and

$$A = (k_b T/h)[\exp(S^{\ddagger}/R)](c^{\circ})^{1-m} \tag{11.82}$$

With the standard state concentration included, Eq. (11.82) gives an entropy of activation for the reaction with the proper units.

5. ENZYME KINETICS

Because many kinetic phenomena within membranes and membrane channels involve an interaction between some species and the membrane, these phenomena are similar to those observed in enzyme-catalyzed reactions in which some substrate S must interact with an enzyme E to react on an adequate time scale. Because the enzyme kinetics provide a parallel for membrane kinetic processes, some techniques of enzyme kinetics will be reviewed here.

Enzyme-catalyzed reactions are bimolecular because the enzyme and substrate must interact to form a complex before the substrate can evolve to product,

$$E + S \underset{k_{-1}}{\overset{k_1}{\rightleftharpoons}} ES \xrightarrow{k_2} P + E \tag{11.83}$$

This is the Michaelis–Menten kinetic scheme. The rate constants k_1 and k_{-1} describe the forward and reverse rates of complex formation, while k_2 describes the rate of formation of product and the release of enzyme from the complex. Note that the equations contain only two enzyme "states," the free enzyme and the enzyme–substrate complex.

The rate equation for the complex is a combination of first- and second-order rate processes,

$$d[ES]/dt = k_1[E][S] - k_{-1}[ES] - k_2[ES] \tag{11.84}$$

Although it might seem more logical to write an expression for the rate of formation of products rather than ES, determination of the complex concentration facilitates the determination of product formation. Equation (11.84) describes a bimolecular, reversible reaction, which is difficult to solve exactly. For this reason, alternative methods of solution are used.

If sufficient substrate S is available, the enzymatic system will reach a steady state. The rate of formation of complex will equal the rate of formation of product. Under these conditions, the concentration of ES will remain constant because each ES that reacts to form product will be replaced by a newly formed ES complex. Under steady-state conditions,

$$d[ES]/dt = 0 \tag{11.85}$$

and Eq. (11.84) can be solved for [ES],

$$d[ES]/dt = 0 = k_1[E][S] - k_{-1}[ES] - k_2[ES] \tag{11.86}$$

and

$$[ES] = k_1[E][S]/(k_{-1} + k_2) \tag{11.87}$$

Note that Eq. (11.87) gives the concentration [ES] in terms of the free enzyme concentration. It is more convenient to express [ES] in terms of the total enzyme concentration,

$$[E_t] = [E] + [ES] \tag{11.88}$$

Equation (11.88) is used to modify Equation (11.86) to

$$0 = k_1[S]\{[E_t] - [ES]\} - (k_{-1} + k_2)[ES] \tag{11.89}$$

Collecting terms and solving gives

$$[ES] = k_1[S][E_t]/(k_1[S] + k_{-1} + k_2) \tag{11.90}$$

where [ES] is now expressed as some fraction of the total enzyme concentration.

The *rate* of product formation at steady state is now determined from the concentration of [ES],

$$R = \text{rate}(P) = k_2[ES] = k_2 k_1[S][E_t]/(k_1[S] + k_{-1} + k_2) \tag{11.91}$$

Note that both rate constants that appear in the numerator are directed toward the product *P*.

A plot of the rate of product formation against substrate concentration would asymptotically approach a limiting rate at larger substrate concentrations, as shown in Fig. 11.3. The formation of complex depends on the concentrations of both E and S. At high concentrations, the enzyme becomes

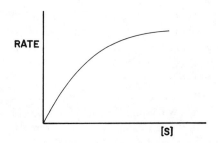

the limiting reagent, that is, there is insufficient enzyme to react with all the substrate present at the high concentrations.

The rate expression can be modified for more convenient graphical examination. The reciprocal of the rate $(1/R)$ is plotted against the reciprocal of substrate concentration $(1/[S])$. The resultant graph is linear. Inverting Eq. (11.91), we find

$$1/R = (k_1[S] + k_2 + k_{-1})/k_1[S][E_t]k_2$$
$$= (1/k_2[E_t]) + \{(k_2 + k_{-1})/k_2 k_1[E_t]\}(1/[S]) \quad (11.92)$$

The inverted concentration $1/[S]$ will approach zero at large substrate concentrations. The intersection of $1/R$ with the ordinate at $1/[S] = 0$ is $1/k_2[E_t]$. Then k_2 is known if the total enzyme concentration is known. This is illustrated in Fig. 11.4.

The slope of this linear plot is equal to a combination of rate constants,

$$\text{Slope} = (k_2 + k_{-1})/k_2 k_1[E_t] \quad (11.93)$$

The slope and intercept permit us to determine only two unknowns. Thus, we cannot determine all three rate constants. The slope is used to determine a ratio called the Michaelis constant,

$$K_m = (k_2 + k_{-1})/k_1 \quad (11.94)$$

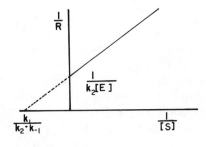

Fig. 11.4 Linearization of Michaelis–Menten kinetics, $1/R$ versus $1/[S]$.

6. GRAPH THEORY KINETICS

The discussion of Michaelis–Menten kinetics in the last section centered on a single enzyme–substrate complex. The concentration of this complex was determined under steady-state conditions and then used to determine the rate of formation of products. Such kinetic problems become more involved when more than one species can bind to the enzyme. For example, if two species compete for the same enzyme site, one of the species may bind strongly and inhibit the reaction of the other species. Alternatively, this inhibitor molecule may bind to an entirely different site on the enzyme while altering the rate of binding of substrate. This might be produced by some binding-induced change in the tertiary structure of the enzyme. It would be most useful to produce some general kinetic method for dealing with such kinetic situations.

Graph theory can be used to develop steady-state solutions for the concentrations of different complexes and free enzyme in solution. To illustrate the method, we reconsider the Michaelis–Menten kinetic scheme,

$$E + S \; \rightleftharpoons \; ES \; \longrightarrow \; P + E \tag{11.95}$$

The total enzyme concentration is divided between two distinct states: the free enzyme and the enzyme substrate complex. The concentrations of these two forms are determined at a steady state using the following rules:

1. A point or vertex is assigned to each distinct enzyme state. For the Michaelis–Menten case, there are vertices for E and ES.
2. If a reaction takes place between two species, the vertices are connected with a line. One line is associated with each reaction process, and an arrow is added to the line to show the direction of that reaction. If the enzyme state can be reached in different ways, an arrow is drawn for each way. The rate constants for each process are now written on their appropriate arrow. The Michaelis–Menten case is shown in Fig. 11.5. The bimolecular rate process of complex formation is approximated by a pseudo-first-order reaction

$$k'[E] = (k_1[S])[E] \tag{11.96}$$

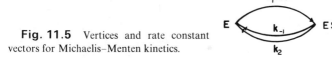

Fig. 11.5 Vertices and rate constant vectors for Michaelis–Menten kinetics.

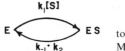

Fig. 11.6 Combination of parallel vectors into a single vector for Michaelis–Menten kinetics.

3. If arrows begin and end at the same vertices, they are combined into a single arrow and their rate constants are added. Fig. 11.6 shows the reduced graph.

4. The fraction of E_t for a given enzyme state is now determined by the ratio of the labelled vectors leading into that state divided by all possible vectors for the system. In this case, the vector $k_1[S]$ leads to complex ES and forms the numerator of our fraction. The sum vector $k_{-1} + k_2$ leads to free enzyme E. The steady-state concentration of ES is then

$$[ES] = [E_t] \{k_1[S]/(k_1[S] + k_{-1} + k_2)\} \qquad (11.97)$$

The vector that leads to E is $k_{-1} + k_2$, so the concentration of E is

$$[E] = [E_t] \{(k_{-1} + k_2)/(k_1[S] + k_{-1} + k_2)\} \qquad (11.98)$$

The sum of these two concentrations is E_t. The expression for ES obtained by this method is identical to that determined in the last section.

5. For multiple-enzyme states, those with more than two vertices, we must find the longest, noncyclic sequence of arrows leading into a given state. The "value" of this sequence is the product of the individual rate constants in a given sequence. If we can find more than one pathway leading to a particular state, each pathway is assigned a separate graph, and the toal vector to describe this state will be the sum of all these graphs. This additional step can be illustrated for a system in which inhibitor I binds only to the enzyme–substrate complex. The kinetic scheme with appropriate rate constants is shown in Fig. 11.7. The total set of vectors are shown in Fig. 11.8. The combinations of vectors that lead to each of the enzyme states and their rate-constant "values" are shown in Fig. 11.9. The product of vectors for ESI is

$$k_1[S]k_3[I] \qquad (11.99)$$

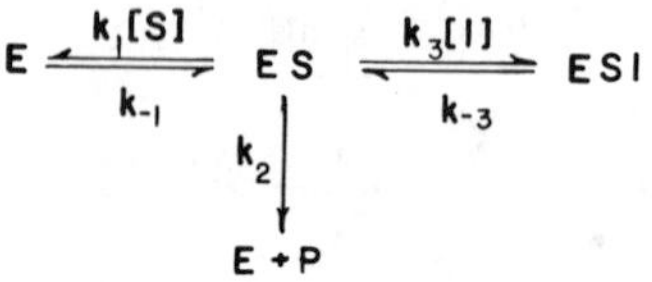

$$E \underset{k_{-1}}{\overset{k_1[S]}{\rightleftharpoons}} ES \underset{k_{-3}}{\overset{k_3[I]}{\rightleftharpoons}} ESI$$

$$\Big\downarrow k_2$$

$$E + P$$

Fig. 11.7 Kinetics for an enzyme–substrate–inhibitor (ESI) system, where S and I bind to independent sites.

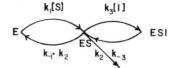

Fig. 11.8 Vectors for the ESI system of Fig. 11.7.

The product for ES is

$$k_1[\text{S}]k_{-3} \qquad (11.100)$$

and the product for E is

$$(k_{-1} + k_2)k_{-3} \qquad (11.101)$$

The steady-state concentration of ES is

$$[\text{ES}] = [\text{E}_t]\{k_1[\text{S}]k_{-3}/(k_1[\text{S}]k_3[\text{I}] + k_1[\text{S}]k_{-3} + (k_{-1} + k_2)k_{-3})\}$$
$$= [\text{E}_t]\, k_1[\text{S}]k_{-3}/\Sigma \qquad (11.102)$$

where Σ is the sum of all the vectors.
The concentration of ESI is

$$[\text{ESI}] = [\text{E}_t]k_1[\text{S}]k_{-3}/\Sigma \qquad (11.103)$$

and the concentration of free enzyme is

$$[\text{E}] = [\text{E}_t](k_{-1} + k_2)k_{-3}/\Sigma \qquad (11.104)$$

Because product is formed only from ES, the rate of product formation is

$$R = k_2[\text{ES}] = k_2 k_1[\text{S}]k_{-3}/\Sigma \qquad (11.105)$$

The rate of product formation depends on E_t and S, as expected. The inhibitor concentration appears explicitly only in Σ, the denominator. For large I concentrations, the proportion of enzyme in the form ESI will increase and lower the fraction of reactive ES.

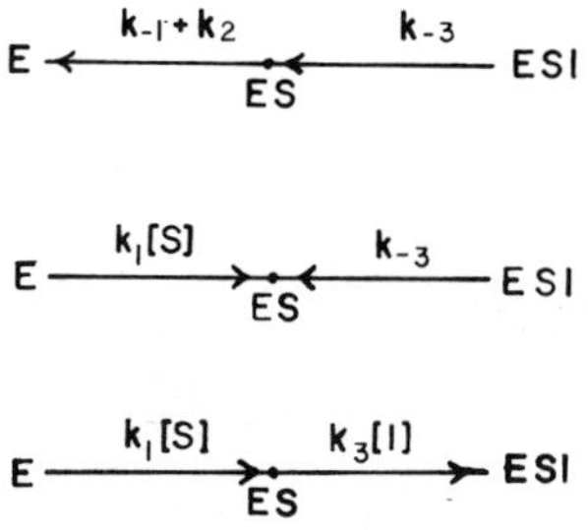

Fig. 11.9 Vector combinations for the fractions of E, ES, and ESI for the ESI system.

7. CYCLIC KINETIC SCHEMES

A cyclic reaction scheme such as the one in Fig. 11.10 raises some interesting kinetic problems. At a steady state, the concentrations of A, B, and C must remain constant, but how restrictive are the conditions that keep them constant? We have suggested previously that the steady state is attained when the rate of appearance of one species, such as A, is exactly balanced by the loss of that species. This implies that the molecules in the system might cycle in some preferred direction. For example, A becomes B, which becomes C, which then becomes A again. If one of the species is removed due to some change in the reaction conditions, then the cycle is broken and the rates for the remaining two species change in a steady state.

The more restrictive condition, which was used by Onsager in his derivation of the reciprocal relations (Chapter 10), suggests that the forward and reverse rates between any pair of species are equal at equilibrium. This is the condition of detailed balance at equilibrium. When detailed balance holds, we could remove the species C without affecting the forward and reverse rates of the A–B reaction. Detailed balance then requires the three kinetic conditions,

$$k_1[A] = k_{-1}[B] \tag{11.106}$$

$$k_2[B] = k_{-2}[C] \tag{11.107}$$

$$k_3[C] = k_{-3}[A] \tag{11.108}$$

To illustrate how these conditions might enter the kinetics of the system, we must determine the steady-state concentrations of the species in the triangular reaction scheme of Fig. 11.10. The graphs for each of the species are illustrated in Fig. 11.11.

The concentration of each species is now described by sums of products, of the form

$$\text{A} \quad k_{-1}k_{-2} + k_3k_{-1} + k_3k_2 \tag{11.109}$$

$$\text{B} \quad k_1k_{-2} + k_1k_3 + k_{-2}k_{-3} \tag{11.110}$$

$$\text{C} \quad k_1k_2 + k_{-1}k_{-3} + k_{-3}k_2 \tag{11.111}$$

with

$$\Sigma = k_{-1}k_{-2} + k_3k_{-1} + k_3k_2 + k_1k_{-2} + k_1k_3 + k_{-2}k_{-3} + k_1k_2$$
$$+ k_{-1}k_{-3} + k_{-3}k_2 \tag{11.112}$$

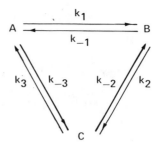

Fig. 11.10 A three-component cyclic reaction system.

When $E = [A] + [B] + [C]$ is the total concentration of all molecules the steady-state concentrations are

$$[A] = E\{k_{-1}k_{-2} + k_3k_{-1} + k_3k_2\}/\Sigma \qquad (11.113)$$

$$[B] = E\{k_1k_{-2} + k_1k_3 + k_{-2}k_{-3}\}/\Sigma \qquad (11.114)$$

$$[C] = E\{k_1k_2 + k_{-1}k_{-3} + k_{-3}k_2\}/\Sigma \qquad (11.115)$$

Because these concentrations are now known, they can be used to determine possible fluxes in the cyclic system. If the system is to satisfy the condition of detailed balance at equilibrium, the flux in the forward direction $(A \rightarrow B \rightarrow C)$ and the flux in the reverse direction must be equal. The net flux from A to B is

$$\begin{aligned}
J_{net} &= k_1[A] - k_{-1}[B] \\
&= (E/\Sigma)k_1(k_{-1}k_{-2} + k_3k_{-1} + k_3k_2) - k_{-1}(k_1k_{-2} + k_1k_3 + k_{-3}k_{-2}) \\
&= (E/\Sigma)(k_1k_{-1}k_{-2} + k_1k_3k_{-1} + k_1k_3k_2 - k_{-1}k_1k_{-2} - k_{-1}k_1k_3 \\
&\quad - k_{-1}k_{-2}k_{-3}) \\
&= (E/\Sigma)(k_1k_2k_3 - k_{-1}k_{-2}k_{-3}) \qquad (11.116)
\end{aligned}$$

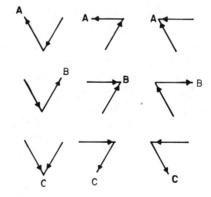

Fig. 11.11 Vector graphs for A, B, and C in the cyclic reaction scheme.

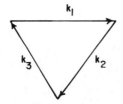

Fig. 11.12 A completed cycle to generate the unidirectional cyclic flow in the A–B–C system.

The forward flux in this case is represented by the product of the forward rate constants, while the reverse flux is represented by a product of the reverse rate constants. When concentrations were determined using graph theory, all vectors led to the state of interest. However, to determine the fluxes, an additional vector must be added to complete a cycle. For example, the flux from A to B is characterized by the product of k_1 and $k_2 k_3$. The vector k_1 completes the cycle (Fig. 11.12). All three vectors move in the same direction.

We will find this same result for each of the three species. Two arrows lead into the state, while the third leaves the state, completing the cycle. The clockwise flow is always

$$[E_t]k_1 k_2 k_3/\Sigma \tag{11.117}$$

while the reverse flow is always

$$[E_t]k_{-1} k_{-2} k_{-3}/\Sigma \tag{11.118}$$

Consider Eq. (11.116) for the net flux. For a zero net flux, the product of the forward rate constants must equal the product of the reverse rate constants,

$$k_1 k_2 k_3 = k_{-1} k_{-2} k_{-3} \tag{11.119}$$

This result, which involves rate constants for all three reactions, is called a Wegscheidner condition. The rate constants for each step may differ, but the product of all forward rate constants must always equal the product of all reverse rate constants. This condition is quite general and does not require that each pair of species must obey detailed balance. Instead, it states that some molecules in the system are cycling clockwise in the reaction scheme while others are cycling counterclockwise, so that the net flux is zero.

The condition of detailed balance is more restrictive. Equations (11.106)–(11.108) can be written as

$$k_1/k_{-1} = [B]/[A] \tag{11.120}$$

$$k_2/k_{-2} = [C]/[B] \tag{11.121}$$

$$k_3/k_{-3} = [A]/[C] \tag{11.122}$$

The product of these three equations is unity,

$$(k_1/k_{-1})(k_2/k_{-2})(k_3/k_{-3}) = ([B]/[A])([C]/[B])([A]/[C]) = 1 \quad (11.123)$$

$$(k_1 k_2 k_3)/(k_{-1} k_{-2} k_{-3}) = 1 \quad (11.124)$$

or

$$k_1 k_2 k_3 = k_{-1} k_{-2} k_{-3} \quad (11.125)$$

Thus, the individual conditions of detailed balance can be combined to produce the Wegscheidner conditions.

The principle of detailed balance does not permit a net flux about some cyclic reaction scheme. The Wegscheidner conditions are more general because they require the product of all rate constants about a cycle to be constant. This suggests that we might be able to induce some transport process by the proper selection of rate constants within a cycle. For this reason, we will abandon the condition of detailed balance for inhomogeneous systems, such as two phases separated by a membrane, and use the constancy of flux around a cycle to determine some membrane transport properties.

8. PASSIVE MEMBRANE TRANSPORT

Graph theory techniques can now be used to describe the transport of ions across membranes. The charged K^+ ion is impermeable in a simple bilayer membrane that is extremely nonpolar. A K^+ concentration gradient cannot produce an observable ionic flow across the membrane. However, the K^+ ion might be complexed with some nonpolar organic species so that the resultant complex becomes membrane-soluble. The complex can then drift across the membrane. If the concentration of K^+ in the opposite bath is lower, the probability of complex dissociation will increase and K^+ will be added to the solution of lower concentration. As this process continues, the concentration of free complexing agent will build on the second side. It will then move back to the original side under this concentration gradient. The cycle is now complete. The agent will be free to complex another K^+ and repeat the cycle.

The structure of a valinomycin molecule is shown in Fig. 11.13. The cavity created in the center of this molecule is the proper size for binding K^+ ion, and the molecule is a selective complexing agent for this ion.

To illustrate the use of cycles in membrane transport, we assume that each of the steps is irreversible. The K^+ transport across the membrane for a unit

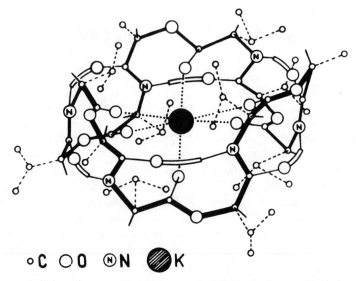

°C ○O ⓃN ⬤K

Fig. 11.13 Structure of the valinomycin-K complex. [Reproduced with kind permission of Ovchinnikov, 1974.]

area of membrane is now shown in Fig. 11.14. The concentration of K^+ on Side 1 is greater than that on Side 2. K^+ from Side 1 forms a complex with free carrier C^- at a rate

$$k_1[C^-][K^+] \tag{11.126}$$

Because we are interested in the four distinct states available to the carrier, which functions as the "enzyme" in these models, the rate constant for this first step is the pseudo-first-order rate constant

$$k' = k_1[K^+]_1 \tag{11.127}$$

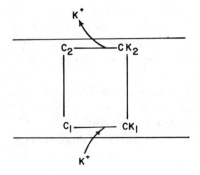

Fig. 11.14 Carrier-mediated passive transport through a membrane: K is absorbed from Bath 1 with concentration c_1, transported, and released to Bath 2 with concentration c_2 ($c_2 < c_1$).

Flux of the complex, is normally determined by the concentration difference of the complex,

$$J = P([KC]_1 - [KC]_2) \tag{11.128}$$

where $[KC]_1$ is the complex concentration near Side 1 and P is the permeability coefficient. For our irreversible case, we assume back diffusion is negligible and the rate of permeation is

$$J = P[KC]_1 \tag{11.129}$$

The dissociation of the complex and release of K^+ occurs at a rate

$$R = k_2[KC]_2 \tag{11.130}$$

where $[KC]_2$, the concentration of complex at the Solution 2 interface, constitutes a state different from $[KC]_1$, the concentration at the Solution 1 interface.

The final step in the scheme is the diffusion of free carrier from Side 2 to Side 1. If the permeability coefficient for this diffusion is P', this final flux is

$$J = P'[C^-]_2 \tag{11.131}$$

As now defined, the two permeation processes are fluxes—that is, moles per second per unit area—while the two reaction processes have the units of moles/volume$^{-1} \cdot s^{-1}$. The reaction rates can be converted to flux units by considering changes in concentration for a membrane interfacial region of thickness λ. With this restriction, all four processes can now be written as fluxes, as shown in Fig. 11.15. The four distinct carrier states form the vertices of this square arrangement.

The graphs used to determine the concentrations of carrier molecules in each of the four states are shown in Fig. 11.16. The concentrations are

$$[C]_1 = [C_t]Pk_2P'/\Sigma \tag{11.132}$$

$$[C]_2 = [C_t]k_1[K]Pk_2/\Sigma \tag{11.133}$$

$$[CK]_1 = [C_t]k_2P'k_1[K]/\Sigma \tag{11.134}$$

$$[CK]_2 = [C_t]P'k_1[K]P/\Sigma \tag{11.135}$$

with

$$\Sigma = Pk_2P' + k_1[K]Pk_2 + k_2P'k_1[K] + P'k_1[K]P \tag{11.136}$$

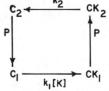

Fig. 11.15 Net fluxes between the four states of the carrier-mediated transport model when all steps are irreversible.

c_2^-

c_1^-

CK_1

CK_2

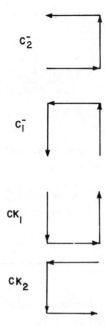

Fig. 11.16 Vector graphs for fractional populations of the four states in the carrier-mediated transport scheme.

The flux for this irreversible cycle is equal to the rate of leaving any one of these carrier states. The rate of leaving $[C]_1$ is

$$J = k_1[K][C]_1 = k_1[K]Pk_2 P'[C_t]/\Sigma \qquad (11.137)$$

The remaining fluxes will also give this result. We have completed the cycle in each case. By analogy with the triangular reaction scheme of the last section, our unidirectional flux is now the product of all four rate constants for the cycle.

Because the unidirectional flux is determined by the product of all four rate constants, the reverse flux in a reversible system is expected to be the product of the four rate constants in the opposite direction. Figure 11.17 gives the rate constants for the reversible scheme. The permeability coeffi-

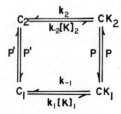

Fig. 11.17 Reversible carrier-mediated passive transport of potassium ion. Forward and reverse permeabilities for each species are equal.

cients P and P' are the same for flow in either direction. For the reverse processes, the bimolecular process with rate constant

$$k_{-2}[K]_2 \tag{11.138}$$

is located on Side 2 while a reversible dissociation k_{-1} is added on Side 1.
    The reverse flux is now proportional to

$$k_{-2}[K]_2 Pk_{-1}P' \tag{11.139}$$

To determine the exact flux relation, we must determine the new denominator, Σ. For the set of rate constants in Fig. 11.17, there are a total of 16 graphs, with 4 graphs for each enzyme or carrier state. The 16 graphs are shown in Fig. 11.18. The 16-term expression for Σ is

$$\begin{aligned}
\Sigma = \ &k_2 Pk_1[K] + P'k_{-1}P + P'k_2k_{-1} + P'Pk_2 \\
&+ k_1[K]Pk_{-2}[K] + P'k_1[K]P + k_{-1}P'k_{-2}[K] + P'k_{-2}[K]P \\
&+ k_1[K]Pk_{-2}[K] + P'k_1[K]P + P'k_2k_1[K] + P'k_{-2}[K]P \\
&+ k_{-2}[K]Pk_{-1} + PP'k_{-1} + P'k_{-1}k_{-2} + P'Pk_2
\end{aligned} \tag{11.140}$$

Although some terms are identical because P is used for permeation in both directions, all terms are listed to show the correlation with the 16 distinct graphs.

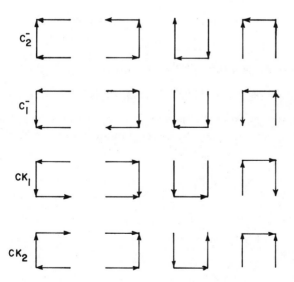

Fig. 11.18 Graphs for each of the four states in the reversible carrier-mediated passive transport model.

9. ION TRANSPORT THROUGH CHANNELS

The methods developed for enzyme kinetics can be extended to a study of ion transport through sites in a channel with very little modification. Consider the simple case of a channel with a single internal binding site, as illustrated in Fig. 11.19. The channel can now be viewed as an enzyme with a single binding site. As such, it can exist in only two distinct configurations: it may be ion-free or have a single ion at the binding site. The system resembles Michaelis–Menten kinetics and will differ only in the way we define the rate constants.

Assume a system in which the total K^+ concentration is found on Side 1. The rate constant for the transition from an empty to a full channel is then

$$k_1[K] \tag{11.141}$$

Once the ion is bound, the empty channel can be regenerated in two ways. The ion may continue to Bath 2 with rate constant k_2, or it may return to Bath 1 with rate constant k_{-1}. The actual flux to Side 2 is then

$$J = k_2[EK] \tag{11.142}$$

The graphs for E and EK are shown in Fig. 11.20. This figure is identical to that for Michaelis–Menten kinetics. The concentrations of E and EK are

$$[E] = [E_t](k_2 + k_{-1})/(k_1[K] + k_2 + k_{-1}) \tag{11.143}$$

$$[EK] = [E_t]k_1[K]/(k_1[K] + k_2 + k_{-1}) \tag{11.144}$$

and the flux of K^+ ion through the membrane is

$$J = k_2[EK] = [E_t]k_2 k_1[K]/(k_1[K] + k_2 + k_{-1}) \tag{11.145}$$

In this case, the "product" is the ion that appears at Side 2. Thus, the only difference between this one-state channel model and the Michaelis–Menten

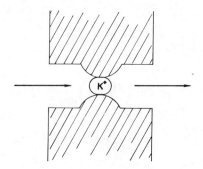

Fig. 11.19 Channel with a cation binding site:

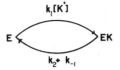

Fig. 11.20 Vector graph for a channel with a single binding site. Only Bath 1 contributes cations to the site.

kinetics lies in the definition of product; $[E_t]$ represents the total number of channels found in some unit area of membrane.

For such reaction site models, we cannot establish a direct proportionality between the flux at the K^+ concentration. The K^+ concentration appears in both the numerator and denominator. The direct proportionality is approached when

$$(k_2 + k_{-1}) \gg k_1[K] \tag{11.146}$$

For such conditions, the rate-determining step is the entry of the ion into the channel. Once the ion does bind, it dissociates to either bath very rapidly.

When K^+ ion is present on both sides of the membrane, the reaction scheme must be modified as shown in Fig. 11.21. The empty site can be filled by two rate processes. The rate of filling by ions from Side 1 is $k_1[K]_1$, while $k_{-2}[K]_2$ is the rate of filling by ions from Side 2. The concentrations of E and EK are

$$[E] = [E_t](k_2 + k_{-1})/(k_2 + k_{-1} + k_1[K]_1 + k_{-2}[K]_2) \tag{11.147}$$

and

$$[EK] = [E_t](k_1[K]_1 + k_{-2}[K]_2)/(k_2 + k_{-1} + k_1[K]_1 + k_{-2}[K]_2) \tag{11.148}$$

The flux to Side 2 will equal the ions leaving the site for Side 2 less those ions that enter the channels from Side 2,

$$J_2 = k_2[EK] - k_{-2}[K]_2[E] \tag{11.149}$$

$$= [E_t](k_2 k_1[K]_1 + k_2 k_{-2}[K]_2 - k_{-2}[K]_2 k_2 - k_{-2}[K]_2 k_{-1})/\Sigma \tag{11.150}$$

$$= [E_t](k_2 k_1[K]_1 - k_{-2}[K]_2 k_{-1})/\Sigma \tag{11.151}$$

with

$$\Sigma = k_2 + k_{-1} + k_1[K]_1 + k_{-2}[K]_2 \tag{11.152}$$

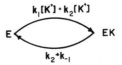

Fig. 11.21 Vector graph for a single-site channel that receives ions from both Bath 1 and Bath 2.

The equation can be simplified considerably if the membrane site is symmetric with respect to both sides. In this case,

$$k_{-1} = k_2 = k' \tag{11.153}$$

$$k_1 = k_{-2} = k \tag{11.154}$$

and

$$J_2 = kk'([K]_1 - [K]_2)[E_t]/\Sigma' \tag{11.155}$$

The flux through the membrane is proportional to the concentration difference across the membrane. However, note again that Σ' contains terms with the two K^+ bath concentrations.

If Eqs. (11.151) and (11.155) describe the steady-state flux into Bath 2, then the rate of flow from Bath 1 into the membrane must be identical. This net flux is calculated using the expression

$$J_1 = k_1[K]_1[E] - k_{-1}[EK] \tag{11.156}$$

Verification of the equality of J_1 and J_2 [Eq. (11.155)] is left as an excercise for the reader.

10. ION TRANSPORT IN MULTISITE CHANNELS

Although we selected a single-site channel to show the strong parallel between enzyme kinetics and ion transport in channels, this model is a minimal description of the channel systems. For example, a more plausible elementary model might be one in which there are binding sites near each interface, as shown in Fig. 11.22. Such models have been proposed for the gramicidin-A channel.

For a two-site model, a total of three carrier states is possible: free channel, channel with an ion at Site 1, and channel with an ion at Site 2. We assume a maximum of one ion per channel at any instant, although this restriction will be removed in a later example.

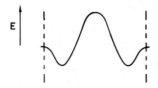

E

Fig. 11.22 A channel with binding sites at each interface and a central energy barrier.

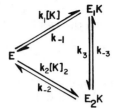

Fig. 11.23 Vector graph for fractional populations of states for the two-binding-site model. Only one K^+ is in the channel at any time.

The full graph with associated rate constants is illustrated in Fig. 11.23, while the two-vector graphs for each species are illustrated in Fig. 11.24. There are now two pathways for a return to E, the empty channel configuration. The ion either leaves Site 1 for Bath 1 or Site 2 for Bath 2. The concentrations for each of the configurations are

$$[E] = [E_t](k_{-1}k_{-2} + k_3k_{-2} + k_{-1}k_{-3})/\Sigma \tag{11.157}$$

$$[E_1K] = [E_t](k_{-2}k_1[K]_1 + k_2[K]_2k_{-3} + k_1[K]_1k_{-3})/\Sigma \tag{11.158}$$

$$[E_2K] = [E_t](k_{-1}k_2[K]_2 + k_3k_2[K]_2 + k_1[K]_1k_3)/\Sigma \tag{11.159}$$

with

$$\Sigma = k_{-1}k_{-2} + k_3k_{-2} + k_{-1}k_{-3} + k_{-2}k_1[K]_1 + k_2[K]_2k_{-3} \tag{11.160}$$

$$+ k_1[K]_1k_{-3} + k_{-1}k_2[K]_2 + k_3k_2[K]_2 + k_1[K]_1k_3$$

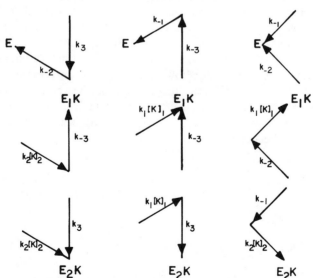

Fig. 11.24 The graphs to determine fractional populations of the three distinct channel states.

Because the net flux must be the same everywhere in the channel, we define the flux using the two bound configurations as

$$J = k_3[E_1K] - k_{-3}[E_2K] \qquad (11.161)$$

The actual expression is generated by substituting the expressions for E_1K (Eq. (11.158)] and E_2K [Eq. (11.159)] into Eq. (11.161).

The rapid increase in the complexity of the expressions as the number of channel sites increases is obvious. However, the graphical method permits us to bypass the complicated rate expressions that would be required for each channel configuration.

Graph-theory methods permit us to examine an additional problem that arises with these multisite channels. Although the ion–ion repulsion produced by two ions in the same channel is expected to be quite large, some of the ionic charge may be neutralized by the binding sites. In this case, we must add an additional state describing two ions at the two binding sites within the channel.

The total graph is quite complicated even for the two-site model, as shown in Fig. 11.25. The $E_{11}K_2$ state is reached when an ion from Bath 1 reaches Site 1 and then an ion from Bath 2 reaches Site 2. It may also be reached in reverse order: Site 2 fills and then Site 1 is filled. We have assumed that the binding at one site does not affect the kinetics at the other site, so the rate constants k_1 and k_2 are retained for each binding process.

Figure 11.25 illustrates a new problem that arises as the number of different configurations increases. To form the different vector graphs, all vectors must lead to the configuration of interest. If we select a sequence of four vectors, we must form a cycle somewhere in the sequence, as shown in Fig. 11.26. Because cycles are not allowed, we can use only three vectors per graph. The number of different three-vector pathways increases for this case. The vector arrangements are shown in Fig. 11.27. A total of six different vector configurations of three vectors each are required for each configuration. Thus, 24 different graphs are required.

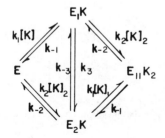

Fig. 11.25 Vector graph for a two-site channel model that permits simultaneous occupation of both sites.

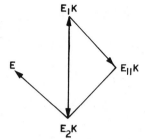

Fig. 11.26 Cycle formation for four vectors in the simultaneous occupation model. A completed cycle such as this cannot be used to determine fractional populations.

The concentrations of the four states are:

$$[E] = [E_t](k_1[K]_1 k_{-2} k_{-1} + k_{-2} k_{-1} k_2[K]_2 + k_3 k_{-2} k_{-1} + k_{-1} k_{-3} k_{-1}$$
$$+ k_{-1} k_{-2} k_{-3} + k_{-2} k_3 k_{-2})/\Sigma \tag{11.162}$$

$$[E_1 K] = [E_t](k_1[K]_1 k_{-2} k_{-1} + k_2[K]_2 k_1[K]_1 k_2 + k_2[K]_2 k_{-1} k_{-3}$$
$$+ k_1[K]_1 k_{-3} k_{-2} + k_1[K]_1 k_{-3} k_{-1} + k_2[K]_2 k_{-3} k_{-2})/\Sigma \tag{11.163}$$

$$[E_2 K] = [E_t](k_1[K]_1 k_2[K]_2 k_{-1} + k_2[K]_2 k_{-1} k_2[K]_2 + k_2[K]_2 k_3 k_{-1}$$
$$+ k_1[K]_1 k_3 k_{-2} + k_1[K]_1 k_3 k_{-1} + k_2[K]_2 k_3 k_{-2})/\Sigma \tag{11.164}$$

$$[E_{11} K_2] = [E_t](k_1[K]_1 k_2[K]_2 + k_2[K]_2 k_1[K]_1 k_2[K]_2 + k_2[K]_2 k_1[K]_1 k_3$$
$$+ k_1[K]_1 k_2[K]_2 k_{-3} + k_1[K]_1 k_3 k_1[K]_1$$
$$+ k_2[K]_2 k_{-3} k_2[K]_2)/\Sigma \tag{11.165}$$

The net flux from the membrane to Bath 2 must include contributions from both the $E_2 K$ and $E_{11} K_2$ states, because these states communicate with Bath 2. The net flux is

$$J_2 = \{k_{-2}[E_2 K] + k_{-2}[E_{11} K_2]\} - \{k_2[K]_2[E] + k_2[K]_2[E_1 K]\} \tag{11.166}$$

The net flux is simply the flux out of the membrane to Bath 2 (the first bracketed term) less the flux into the membrane from Bath 2 (the second bracketed term).

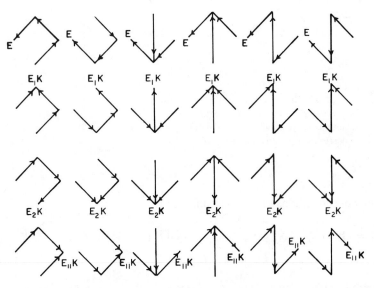

Fig. 11.27 Graphs for fractional populations of states in the simultaneous occupation model.

11. NON–STEADY-STATE GRAPH KINETICS

Although graph theory has been used to establish steady-state concentrations and fluxes for some very complicated systems, it has been limited by the fact that these results are steady-state. However, with a slight modification of the basic approach, we can extend these ideas to systems that change with time.

Consider the basic Michaelis–Menten scheme once again,

$$E + S \underset{k_{-1}}{\overset{k_1}{\rightleftharpoons}} ES \overset{k_2}{\longrightarrow} E + \text{products} \qquad (11.167)$$

The kinetic equation for ES is

$$d[ES]/dt = k_1[S][E_t] - (k_1[S] + k_2 + k_{-1})[ES] \qquad (11.168)$$

In order to solve this equation graphically, we introduced the steady-state approximation,

$$d[ES]/dt = 0 \qquad (11.169)$$

which converted the equation from a differential equation to an algebraic equation. The general information on the time dependence of the reaction was lost in the process.

It is possible to convert our differential equation into an algebraic one without sacrificing the rate information. The equation can be Laplace transformed to create an algebraic equation that can be solved using graph theory methods.

The Laplace transform of some $f(t)$ is

$$L(p) = \int_0^\infty f(t) \exp(-pt)\, dt \qquad (11.170)$$

The definite integral will remove all t dependence from the expression.

The right side of Eq. (11.168) does not contain time explicitly. The first term will move in front of the integral, and we must evaluate

$$k_1[S][E_t] \int_0^\infty \exp(-pt)\, dt \qquad (11.171)$$

Thus,

$$-Cp^{-1}[\exp(-pt)]\Big|_0^\infty = Cp^{-1} \qquad (11.172)$$

where

$$C = k_1[S][E_t] \qquad (11.173)$$

The substrate concentration is assumed to remain constant during the entire kinetic approach to a steady state.

The second term on the right involves ES, which is expected to change with time. Therefore, we must take its Laplace transform. Because the leading rate constants including $k_1[S]$ are assumed constant, they move in front of the integral,

$$(k_1[S] + k_{-1} + k_2) \int_0^\infty [ES] \exp(-pt)\, dt = DF(p) \qquad (11.174)$$

where D defines the sum of constants and $F(p)$ is the Laplace transform of ES.

The Laplace transform of the rate term on the left-hand side must be evaluated using integration by parts,

$$\int_0^\infty (d[ES]/dt) \exp(-pt)\, dt = [ES] \exp(-pt)\Big|_0^\infty$$
$$- \int_0^\infty [ES][\exp(-pt)](-p)\, dt \qquad (11.175)$$

If the concentration of ES is zero at the start of the reaction, then both limits of the first expression are zero. The second term is just the Laplace transform of [ES] multiplied by p.

The Michaelis–Menten equation has been transformed to an algebraic equation,

$$pF(p) = Cp^{-1} - DF(p) \qquad (11.176)$$

The Laplace transform is

$$F(p) = Cp^{-1}/(p + D) \tag{11.177}$$

If we replace the original rate constants, this equation is

$$F(p) = k_1[S][E_t]/(k_1[S] + p + k_{-1} + k_{-2}) \tag{11.178}$$

This expression is very similar to the one we generated for the steady state. The factor of p generated by the derivative appears in the denominator as an additional rate constant. Because it does not appear in the numerator, it has the same characteristics as k_{-1} and k_2, that is, leading back to the free enzyme. It can now be included as an additional vector in the transform space. The p^{-1} in the numerator is not associated with $k_1[S]$ in the denominator. This indicates it is connected with E_t. Thus, we can establish rules to include p in the transform space. The p is included as an additional vector leading to the free enzyme, and E_t is changed to the form $p^{-1}E_t$. The new vector diagram is shown in Fig. 11.28.

For more complicated systems, a vector p will lead from each bound enzyme state to the free enzyme state. In all these cases, we must assume that all enzyme is present in the free state at time zero.

Once we have established the form of the Laplace transform, it must be restored to a time-dependent expression via an inverse Laplace transform. Equation (11.177) can be rewritten as

$$F(p) = C/(p + 0)(p + D) \tag{11.179a}$$

to conform to an expression of the form

$$F(p) = 1/(p + A)(p + D) \tag{11.179b}$$

with inverse Laplace transform

$$[\exp(-At) - \exp(-Dt)]/(D - A) \tag{11.180}$$

The inverse Laplace transform for Eq. (11.192) is

$$ES(t) = k_1[S][E_t][\exp(0) - \exp(-Dt)]/(D - 0) \tag{11.181}$$

$$= k_1[S][E_t](1 - \exp\{-(k_1[S] + k_{-1} + k_2)t\})/(k_1[S] + k_{-1} + k_2) \tag{11.182}$$

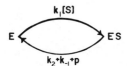

Fig. 11.28 Michaelis–Menten kinetics vector graph in transform space. The vector p carries the time-dependent information for this model.

The decay constant is just the sum of all the rate constants. For long times, the exponential term decays to zero and we observe the steady-state result,

$$[ES] = k_1[S][E_t]/k_1[S] + k_{-1} + k_2) \qquad (11.183)$$

PROBLEMS

1. A substrate S reacts to form products P. The enzyme for this reaction must be activated by the molecule A, where A and S bind to separate sites of the enzyme E. Calculate the rate of formation of products using graph theory.

2. Determine the time-dependent unidirectional flow of a positive ion through a membrane having a single charge-binding site. The input and output fluxes must be equal at steady state. Show that this is correct.

3. A simple carrier system is defined as a molecule that has two configurations,

$$R \underset{k_{-1}}{\overset{k_1}{\rightleftharpoons}} T$$

and the configuration T binds substrate from either Bath 1 or 2 with rates $k[S]_1$ and $k[S]_2$, respectively. The constant can dissociate in either direction, with rate constants k_2 for each direction. (a) Draw the graphs for steady-state concentrations of R, T, and TS. (b) Find the steady-state TS concentration. (c) Determine the net flux of substrate.

4. An enzyme E can function as a carrier by accepting an ion from one side of a membrane to form a complex and then dissociating to release the ion to the other side of the membrane. (a) Set up the graphs for this process. (b) Find the steady-state concentration of ES. (c) Show that the fluxes into and out of the membrane are equal.

Chapter 12 _____

Membrane Excitability

1. THE ACTION POTENTIAL REVISITED

For biological excitable membranes, the phenomenon of excitation most commonly studied involves a sudden increase in membrane permeability to Na^+ ions. The electrochemical potential due to these Na^+ ions produces a net reversal in the transmembrane potential. The sodium permeability then decreases and the membrane returns to its equilibrium potential, which is dominated by the K^+ ion concentrations.

This may appear to be adequate information, because it does explain the transient behavior of the excitation process. However, on closer examination, it raises more questions than it answers. For example, what type of molecular mechanism is responsible for the transient increase in sodium permeability? How do the ions pass through the membrane? Do the K^+ and Na^+ ions follow the same pathways through the membrane? To answer such questions, a detailed molecular picture of the whole excitation mechanism at the molecular level is required and, for this reason, we must address some of the techniques that can be used to deduce this mechanism.

2. THE VOLTAGE CLAMP

In Chapter 6, the voltage clamp was introduced. In such a system, the potential across the membrane is maintained at a potential equivalent to some control potential by placing the axon in the feedback loop of an operational amplifier circuit. Thus, the situation differs significantly from that observed with the stimulated action potential. For the action potential, the only net current that flows through the membrane is the externally applied stimulating current. In the voltage clamp system, the operational amplifier must provide sufficient current through the membrane to maintain the transmembrane potential, and current may flow during the entire voltage clamp process. This does not complicate the problem, however, because the experimenter has complete control over the potential which he applies. For example, a control step pulse of constant potential could be applied, and this would maintain a potential. transmembrane. The experimenter would then observe the current which flowed through the membrane at this constant potential. If this current changes with time while the potential remains constant, the resistance (or conductance) of the nerve membrane is changing. This is consistent with our earlier observation that the permeability of Na$^+$ ion increases during excitation. If Na$^+$ ion is also capable of carrying current, the total amount of current that can move through the membrane must increase.

The currents observed during a voltage clamp for a series of constant potential steps are shown in Fig. 12.1. There are two major current regions. During the initial phase, current flows from the exterior to the interior of the membrane. This negative current is an "inward" current. After a period of time, the inward current phase disappears and is replaced by an outward current phase, which rises to a steady current level. The larger currents observed in this steady state suggest that the membrane remains more

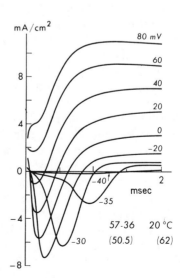

Fig. 12.1 Typical voltage clamp results: membrane current densities for clamp potentials as labelled. [Reproduced with kind permission of Cole, 1968].

permeable to ions at the applied potential and permits a large steady flow of ions from the interior to the exterior of the cell.

From previous discussions, the inward current is perceived as a flow of of Na^+ ions. The large excess of Na^+ ions in the external solutions makes this ion the most likely charge carrier. However, how can this be proved? The most logical approach involves removing Na^+ ions from the external solution and determining any changes in the magnitude of the internal current. The Na^+ ion must be replaced by a second ion — such as choline — so that the net internal and external ion concentrations remain equal. If there were an ionic imbalance, the resultant osmotic pressure flow could produce a volume change in the cell, which would destroy its function. When the Na^+ concentration is lowered, the maximum of the inward current during a voltage clamp also falls and disappears as the concentration of Na^+ approaches zero. The bulk of the inward current is apparently carried by the Na^+ ions.

The outward current is carried by the K^+ ions, and this suggests that the net K^+ flow will be maintained at a steady-state level as long as the clamping potential is maintained. The mechanism that permits the enhanced transport of K^+ ion is altered by the applied potential, and a maintained potential difference leads to an enhanced membrane current. The mechanism that permits this enhanced flow remains "on" as long as the potential difference is maintained.

This raises an additional question. Are the flows of Na^+ and K^+ ion coupled in some way? In other words, does the inward flow of Na^+ ion trigger some mechanism, which then enhances the outward flow of K^+ ion? What methods might be used to determine whether such coupling exists? It is possible to eliminate either the Na^+ or K^+ currents from the record without disturbing the other current. For example, if the puffer fish toxin, tetrodotoxin, is added to the external solution in nanomolar quantities, the sodium current component can be completely eliminated. However, there is no concomitant change in the K^+ current. This suggests a complete independence of Na^+ and K^+ pathways through the membrane. If TEA^+ is added to the cell interior, the K^+ currents can be eliminated while the Na^+ currents remain. These chemical blocking agents provide evidence that the Na^+ and K^+ channels are separate and uncoupled. The independence of the channel currents can also be established by inducing changes in the kinetics of the excitation process. If the systems are coupled, changes in the kinetics of the two processes should be correlated. However, no such correlation is observed.

The currents through the membrane are produced by ions that flow through channels that traverse the membrane. Independent ensembles of channels exist for both the Na^+ and K^+ ions. The alternate mode of transport, the use of carriers, could produce saturation. If large numbers of ions must flow through the membrane during a voltage clamp experiment, the available carriers might then be unable to handle the flow and the current would reach a limiting value. Although some current limits are observed in squid giant axon, these are due to alternate causes, which will be discussed below.

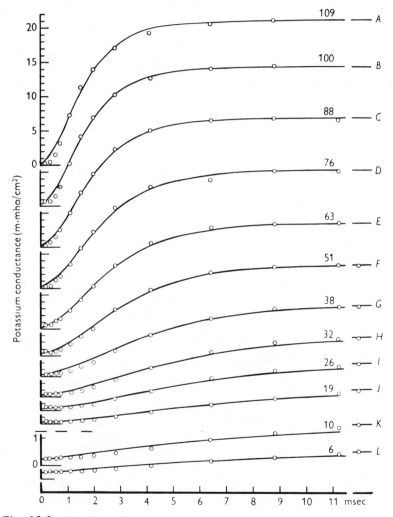

Fig. 12.2 Rise of potassium conductance associated with different depolarizations. The circles are experimental points obtained on axon 17, temperature 6–7°C, using observations in sea water and choline sea water. [Reproduced with permission from Hodgkin, 1952.]

If the ions do move through independent channels in the membrane, it then becomes easier to examine some qualitative aspects of their kinetics. In Figs. 12.2 and 12.3, a series of current records for K^+ and Na^+ ions at a series of clamping potentials are shown. In Fig. 12.2, the K^+ currents rise to progressively larger steady-state current levels as the magnitude of the clamping potential is increased. In addition, the rise to this steady state takes place more rapidly when the clamping potential is increased. A kinetic scheme that describes such currents must provide a consistent explanation for both these phenomena.

The sodium currents of Fig. 12.3 for a series of clamping potentials are more complicated. As the potential is increased, the initial phase of the inward sodium current falls progressively more rapidly. However, the magnitude of the maximal inward current simultaneously decreases and finally

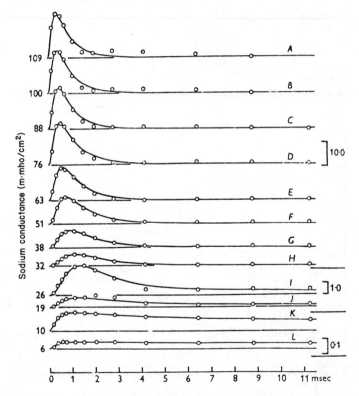

Fig. 12.3 Changes of sodium conductance associated with different depolarizations. The circles are experimental estimates of sodium conductance obtained on axon 17, temperature 6–7°C. [Reproduced with permission from Hodgkin, 1952.]

becomes a net outward current (with a fast rise time) when the applied clamping potential is approximately 130 mV or larger. The speed with which the current "turns off" also increases with increasing membrane potential. A consistent model is necessary to describe all these observations.

3. THE HODGKIN–HUXLEY EQUATIONS—PRELIMINARY CONSIDERATIONS

To develop a consistent model for the current records of the last section, additional experimental information is needed to limit the number of possible descriptions necessary for the system. A central experiment of this type is illustrated in Fig. 12.4. The clamping potential is applied across the membrane by applying a step potential pulse of finite duration to the input of the voltage clamp system. When this pulse is first applied, a short capacitative current transient is observed due to the intrinsic capacitance of the nerve membrane. At the end of the step pulse, when the potential returns to its original value, a negative capacitance transient is observed, as expected. However, the current does not return instantaneously to its original zero current value. The decay to zero current is exponential and requires time in the millisecond range. The capacitative transient is finished in <0.1 ms. In other words, the membrane is still conducting current even though the driving potential has been eliminated.

This basic observation can be extended by jumping the step pulse to a series of potentials rather than the single zero potential pulse. (Note that a "zero" potential is actually the −60 mV resting potential of the membrane, which constitutes the equilibrium value of the unexcited membrane.) As the potential is made progressively more negative—for example, −80 mV, −100 mV—the current jumps to more negative current levels and the subsequent decay becomes faster. If the voltage is jumped to a value higher than −60 mV, such as −40 mV, the magnitude of the current jump decreases and the decay to zero current becomes progressively slower.

To understand the significance of such experiments, this jump data is plotted on a current-versus-voltage curve (Fig. 12.5). The initial potential step has an absolute potential of +20 mV (80 mV above the −60 mV resting potential) and produces a current of 3 mA/cm² at this level. This is plotted as one point on the graph. The other points are determined from the *final* absolute potential and the current that is observed for this potential immediately after the jump, so that no decay is possible. The capacitative

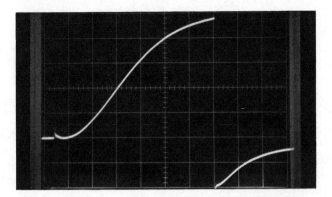

Fig. 12.4 A repolarization experiment in which an axon clamped at voltage V is suddenly clamped at $V = 0$.

transient is eliminated before determining the maximum current. Alternately, subsequent current points may be extrapolated back to the $t = 0$ point to establish the initial current. For example, the jump to -50 mV produces an initial current of 0 mA/cm^2, and this point is plotted on the graph.

When all the points are plotted in this manner, the result is a linear current-versus-voltage plot. This suggests that the membrane functions ohmically, as Ohm's law states that current is proportional to potential,

$$I = g(V - V_{\mathrm{o}}) \tag{12.1}$$

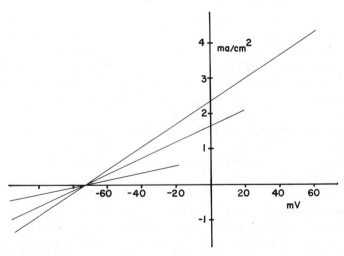

Fig. 12.5 Instantaneous current-versus-voltage curves. The instantaneous current is observed immediately after the repolarization.

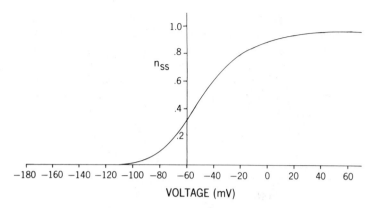

Fig. 12.6 Steady-state current versus absolute clamping potential in squid giant axon, illustrating the sigmoidal dependence on potential.

where I is a current density or charge flux (the current through a unit area of membrane), g is the conductance of a unit area of membrane, V is the potential, and V_o is present to reflect the fact that the current becomes zero when the net membrane potential is $-72\,\text{mV}$, not $0\,\text{mV}$. If V_o is $-72\,\text{mV}$, no current is observed when the membrane is jumped to this potential.

Figure 12.6 illustrates curves of I versus V for a series of different potential steps. Once again, the curves of I versus V are linear, but their slopes now differ. As the initial step potential increases, the slope of the lines also increases. The conductance g is related to this slope because the slope is

$$g = I/(V - V_o) \qquad (12.2)$$

for this linear plot, so these data suggest that the conductivity of the membrane is increasing as the clamping potential increases. In other words, the membrane permits a larger ion flow at the larger potentials. Because the linear ohmic relationship reflects the electrical properties of the membrane at the instant of jump, the data suggest that the membrane obeys Ohm's law during this short period of time (the time required to jump from one potential to another).

Although this might suggest that the membrane obeys Ohm's law continuously, this is not the case. Ohm's law is obeyed during the potential "jumps." Figure 12.6 shows a current-versus-voltage curve for steady-state currents as a function of voltage. This plot has a definite curvature; it rises very slowly for the smaller potentials and then rises very rapidly as the potential is increased. The curve of I versus V is similar to that observed for a diode. The diode (or the nerve membrane) conducts when the potential exceeds some value. Below that potential, the current is minimal.

The differences between the instantaneous and steady-state current-versus-voltage curves can be reconciled if we remember that the steady-state conductance g changed as the clamping potential changed. This suggests that during a voltage clamp experiment, the membrane conductance is changing with time. If a larger potential is applied, the conductance will attain a larger steady-state conductance at a more rapid rate. Ohm's law now takes the form

$$I(t) = [g(t)](V - V_o) \qquad (12.3)$$

The current rises during a voltage clamp experiment because the conductance is changing with time. The shape of the temporal current record is then dictated by the kinetics of the conductance changes.

Equation (12.3) also provides a consistent explanation for the linear instantaneous current–voltage relations. During the rapid jump between potentials, the kinetic processes for the conductance have insufficient time to change and the conductance appears to be essentially constant over the interval. The subsequent decay of the current reflects the slower kinetic response of the conductance changes.

The decay of current after the jump is exponential in all cases, and this suggests that some first-order kinetic process may be present. However, such an exponential decay must then be correlated with the distinctive sigmoidal rise of current when the initial step potential pulse is applied.

4. THE HODGKIN–HUXLEY K⁺ EQUATION

Because the K currents are now assumed to obey an equation of the form of Eq. (12.3), it becomes possible to focus on the time dependence of the conductance. The exponential decay observed after a potential jump indicates a first-order process, and this was used as a starting point by Hodgkin and Huxley. Hodgkin and Huxley postulated gating molecules within or near the K^+ channels that obeyed first-order kinetics. These gating molecules could exist in either a closed or an open configuration; the closed configuration retarded current flow in the channel, while the open configuration permitted such flow. If n is defined as the fraction of such gating molecules that are in the open configuration, then the fraction in the closed configuration must be $1 - n$. The transitions between closed and open configurations are now described by first-order kinetics. The equation is

$$dn/dt = \alpha(1 - n) - \beta n = \alpha - (\alpha + \beta)n \qquad (12.4)$$

where the rate constants α and β depend on the potential applied to the membrane.

This equation can now be solved to determine $n(t)$. The integrating factor for this first-order differential equation is

$$\exp[+(\alpha + \beta)t] \qquad (12.5)$$

and the equation can be rewritten as

$$\exp[-(\alpha + \beta)t][dn/dt + (\alpha + \beta)n] = \alpha \exp[(\alpha + \beta)t] \qquad (12.6a)$$

and the left-hand side can be compressed as

$$d/dt[n(t)\exp(\alpha + \beta)t] = \alpha \exp[(\alpha + \beta)t] \qquad (12.6b)$$

If the fraction of open channels is n_o at time zero and $n(t)$ at time t, the equation can be solved by integration:

$$\int_{n_o}^{n} d[n(t)\exp(\alpha + \beta)t] = \int_{n_o}^{t} \exp[(\alpha + \beta)t]\,dt \qquad (12.7)$$

$$= n(t)\exp[(\alpha + \beta)t] - n_o$$

$$= \{\exp[(\alpha + \beta)t] - 1\}/(\alpha + \beta) \qquad (12.8)$$

Eliminating the exponential from the left-hand side and shifting n_o leads to the final form for $n(t)$:

$$n(t) = n_o \exp[-(\alpha + \beta)t] + [\alpha/(\alpha + \beta)][1 - \exp\{-(\alpha + \beta)t\}] \qquad (12.9)$$

This equation can be simplified still further by considering the steady-state behavior of the equation. When $dn/dt = 0$ after application of the clamping potential, the equation is

$$0 = \alpha - (\alpha + \beta)n \qquad (12.10)$$

$$n_\infty = \alpha/(\alpha + \beta) \qquad (12.11)$$

When this steady-state value is substituted into Eq. (12.9), it becomes

$$n(t) = \{n_\infty - (n_\infty - n_o)\exp[-(\alpha + \beta)t]\} \qquad (12.12)$$

This solution describes an exponential rise of the fraction of open configurations and does not describe the sigmoidal shape of the voltage clamp current records which are observed for squid giant axon. However, when the membrane was repolarized—that is, returned to a lower clamping potential from its initial clamping potential—an exponential decay to zero current was observed. For the repolarization process, the initial fraction of open molecules, n_o must be large while the final steady-state value of such molecules at the final lower potential must be small, that is, $n_\infty \simeq 0$. Equation (12.12) is then approximately

$$n(t) = n_o \exp[-(\alpha + \beta)t] \qquad (12.13)$$

and the decay appears exponential, as actually observed.

How can this exponential form be reconciled with the sigmoidal form observed when the final potential (and n_∞) is larger than the initial potential? Hodgkin and Huxley noted that the exponential in Eq. (12.13) could be raised to the fourth power without altering the essential form of the equation:

$$n(t)^4 = n_o^4 \exp[-4(\alpha + \beta)t] \qquad (12.14)$$

If such a fourth-power dependence is now applied to the general expression for $n(t)$ [Eq. (12.12)], then the fourth-power dependence reproduces the sigmoidal form of the current records when $n_o < n_\infty$.

Since no current is possible if the membrane is clamped to the equilibrium potential, the K^+ equilibrium potential is used to define the zero of current for the K channels. Because the instantaneous response is ohmic, the time-dependent current through an ensemble of K channels is

$$i(t) = \bar{g}_K[n(t)]^4(V - V_K) \qquad (12.15)$$

where V_K is the equilibrium K potential and $\bar{g}_K$ is the total conductance of an ensemble of K channels in 1 cm^{-2} of membrane surface. If all the channels in this area are completely open, $\bar{g}_K$ represents the maximum conductance for the ions through this region, with units of siemens per square centimeter (S/cm^2).

For a given clamping potential, the membrane conductance can only approach this maximum, because the steady-state value of $n(t)$ will depend on the magnitude of this clamping potential. The steady-state fraction of open molecules,

$$\alpha/(\alpha + \beta)$$

will decrease with decreasing clamping potential so that

$$i_{ss} = \bar{g}_K n_\infty^4 \qquad (12.16)$$

This equation can be used to reproduce the curve of steady-state current versus potential (Fig. 12.6).

The equations developed here can be used to determine the temporal development of the current for any clamping potential. However, the rate constants α and β both depend on this clamping potential. Hodgkin and Huxley determined values of these rate constants for each potential and then developed empirical relations that could be used to determine these rate constants for any potential. The expressions are

$$\alpha_n = -0.01(V + 50)/\{\exp[0.1(V + 50)] - 1\} \qquad (12.17)$$

$$\beta_n = 0.125 \exp[(V + 60)/80] \qquad (12.18)$$

where the subscript n signifies K channel rate constants. The basic shapes of these functions are shown in Fig. 12.7. The rate constant α_n that dominates the kinetics at the larger depolarizing potentials becomes linear in this potential range, because the denominator approaches 1.

The functional form of α_n was chosen in analogy to the form of the Goldman equation, which contains a similar exponential term in its denominator. However, Eqs. (12.17) and (12.18) are empirical equations to describe the experimental data.

When the membrane is clamped to a potential that is more positive than its resting equilibrium potential, this potential is called a depolarizing potential and the process is a depolarization. When the membrane is clamped to a potential more negative than its resting potential, such as -100 mV, the clamping potential is a hyperpolarizing potential. When the clamping potential is switched from one clamping potential to a second clamping potential, as was done in the instantaneous current–voltage experiments,

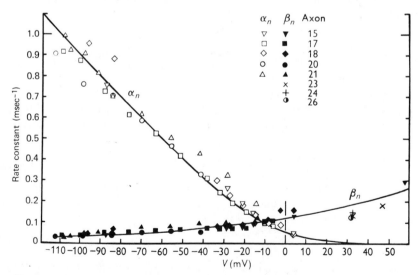

Fig. 12.7 Abscissa: membrane potential minus resting potential in sea water. Ordinate: rate constants determining rise (α_n) or fall (β_n) of potassium conductance at 6°C. [Reproduced with permission from Hodgkin, 1952.]

the process is called a repolarization. Generally, -60 mV constitutes the boundary between hyperpolarization and depolarization, so that a potential could be negative, such as -30 mV, and still be a depolarizing potential.

5. THE HODGKIN–HUXLEY Na$^+$ EQUATION

A description of the sodium currents must necessarily be more complicated than the K currents, because the sodium current must turn on and turn off, while the K currents only turn on when a depolarizing potential is applied. Like the K current equations, the Na current equations are assumed to be ohmic since they exhibit similar instantaneous current–voltage responses when the potential is switched while the sodium currents are flowing. The equation is

$$i_{Na}(t) = [g_{Na}(t)](V - V_{Na}) \tag{12.19}$$

where V_{Na} is now the equilibrium potential of the sodium. This V_{Na} is observed in the range of $+50$–60 mV, providing an immediate explanation of the appearance of both inward and outward sodium currents. When the depolarization involves potentials below 60 mV, the equilibrium potential exceeds V_{Na} and is thus capable of driving ions into the cell when the sodium channels are open. This leads to inward current records. As the clamping potential reaches V_{Na}, the sodium current is eliminated entirely because the clamping potential exactly balances the equilibrium sodium potential. When the clamping potential is raised to exceed V_{Na}, current can be driven outward, as observed in the sodium current records.

Now $g_{Na}(t)$ must be designed to include the kinetics of both turn-on and turn-off. The kinetic turn-on process is described by a first-order kinetic process in analogy with the K system. The equation for open molecules, m, is

$$dm/dt = \alpha_m(1 - m) - \beta_m m = \alpha_m - (\alpha_m + \beta_m)m \tag{12.20}$$

where the rate constants α_m and β_m now describe the opening of gating molecules in the sodium channels. This mathematical solution is identical to that for K channels, and the final result is

$$m(t) = \{m_\infty - (m_\infty - m_o)\exp[-(\alpha_m + \beta_m)t]\} \tag{12.21}$$

Hodgkin and Huxley created the sigmoidal form by raising $m(t)$ to the third power,

$$m(t)^3 \tag{12.22}$$

This third-power dependence is sufficient to describe the initial phase of the sodium current. However, a second kinetic process is now required to turn off the transient sodium current. Once again, Hodgkin and Huxley postulated a first-order process for the turn-off. They postulated a fourth, turn-off, molecule for the channel to complement the three turn-on gating molecules. The fraction of turn-off molecules, h, was described by the kinetic equation

$$dh/dt = \alpha_h(1 - h) - \beta_h h = \alpha_h - (\alpha_h + \beta_h)m \qquad (12.23)$$

Using both the turn-on and turn-off factors, the conductance as a function of time for the sodium channels now becomes

$$g_{Na}(t) = \bar{g}_{Na}[m(t)]^3[h(t)] \qquad (12.24)$$

where

$$h(t) = h_\infty - (h_\infty - h_o)\exp[-(\alpha_h + \beta_h)t] \qquad (12.25)$$

and $\bar{g}_{Na}$ is the maximal conductance of the channels in a unit membrane area when all these channels are open.

Hodgkin and Huxley developed empirical relationships for all the sodium channel rate constants as functions of voltage. The m process rate constants are

$$\alpha_m = [0.1(V + 35)]/\{\exp[0.1(V + 35)] - 1\} \qquad (12.26)$$

and

$$\beta_m = 4\exp[(V - 60)/18] \qquad (12.27)$$

The factors in the numerator are approximately 10 times as large as those used for the K$^+$ channel rate constants, and this reflects the significantly increased rate with which the Na$^+$ channels open. This, of course, permits the sodium current transient to occur before there has been appreciable change in the fraction of open potassium channels.

The h process rate constants are

$$\alpha_h = 0.07\exp[(V - 60)/20] \qquad (12.28)$$

and

$$\beta_h = \{\exp[0.1(V + 30)] + 1\}^{-1} \qquad (12.29)$$

The magnitude of these rate constants is similar to that observed for the K$^+$ channel turn-on process. The h process rate constants appear to have a form that differs from the two turn-on processes. The forward rate constant is now the simple exponential function. This may appear quite arbitrary, but it is not. For the *h* process, a large fraction of the *h* gates are assumed to be in their open (nonblocking) configuration when the potential is near equilibrium. During a depolarizing voltage clamp, this large fraction of nonblocking molecule evolves exponentially toward $h = 0$, and this serves to turn off the current through the membrane. Thus, even though the three "on" gating molecules act to open the channel during the voltage clamp, the *h* gating molecule acts to reclose it. Because the turn-on process is faster, a sequence of open followed by closed channel is observed. The combination of the *m* and *h* processes produces the transient Na^+ current illustrated in Fig. 12.8.

In addition to the independent Na^+ and K^+ channel currents, additional current pathways through the membrane also exist. However, these pathways show no kinetic variations on the time scale of the K and Na channels. Because the conductance of these pathways will remain constant throughout the voltage clamp experiment, the conductance is constant and the final leakage component can be expressed through Ohm's law as

$$i_1 = g_L(V - V_L) \tag{12.30}$$

The total current through the membrane during a voltage clamp to a specific voltage is then

$$I(t) = i_K(t) + i_{Na}(t) + i_L \tag{12.31}$$

The parameters for these three membrane currents are:

$$\bar{g}_K = 0.036 \, \text{S/cm}^2 \qquad \bar{g}_{Na} = 0.120 \, \text{S/cm}^2 \qquad g_L = 3 \times 10^{-4} \, \text{S/cm}^2$$

$$V_K = -72 \, \text{mV} \qquad V_{Na} = +55 \, \text{mV} \qquad E_L = -49.4 \, \text{mV}$$

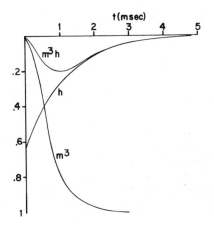

Fig. 12.8 Generation of the transient sodium channel conductance using the function $[m(t)]^3 h$.

Note that if the conductance is expressed in S/cm^2 and the potential is expressed in mV, the resultant current will have the dimensions of mA/cm^2. This set of dimensions is the most common format for this system.

6. EFFECTS OF POLYVALENT IONS

The rate constants for channel kinetics in voltage-clamped axons are definite functions of the transmembrane potential. However, this dependence does not indicate how the potential controls the gating processes. It seems reasonable to assume, however, that the local electric field near the gating molecules controls the changes that lead to channel opening. This local field cannot be probed directly with the transmembrane potential. However, it may be possible to alter this electric field by alternate means, and the most obvious approach is the introduction of polyvalent ions. When polyvalent ions enter the double-layer region the electrical forces generated by these ions can have some effect on molecules within the membrane, particularly if these molecules lie relatively close to the membrane surface. Such ions could then perturb the local electric field of the gating molecules and, because the local field controls the kinetics of the gating process, it is possible to establish a correlation between the ion concentrations and changes in the local electric field.

The ion used frequently in such studies is the calcium ion, because this ion occurs naturally in biological systems and plays a role in stabilizing membrane systems. For example, if divalent ion concentration is lowered in the external solution for squid giant axon, the axon begins to fire spontaneously and repetitively.

The calcium ion concentration for squid giant axon can be varied over a considerable range, for example, $5\,mM$ to $120\,mM$. In this concentration range, Frankenhaeuser and Hodgkin (1957) noted definite changes in the membrane kinetics. By observing changes in the kinetic rate constants as a function of calcium ion concentration for a given clamping potential, they were able to deduce the magnitude of potential shifts induced by the Ca^{2+} ions. They reported that a five-fold reduction in the external Ca^{2+} concentration shifts the kinetic parameters of the squid axon in a manner equivalent to an electrical depolarization of $10-15\,mV$. Similar shifts were observed for the kinetic parameters of both the sodium and potassium channels, and this suggests that the shifts are produced by a homogeneous field distributed over the entire membrane surface.

The magnitude of the potential shifts increases markedly when trivalent ions are substituted for divalent ions. The double-layer potential generated by the trivalents is expected to be significantly larger than that generated by the divalents, and this is indeed observed. Comparable potential shifts of the kinetic parameters are produced by lanthanide ion concentrations in the range of 0.01–0.1 mM. Although this concentration is two orders of magnitude lower than the concentrations required to produce a comparable shift with divalent ions, it is consistent with the Grahame equation because the power of three in the exponential term permits a concomitant reduction in the concentration to produce a comparable effect.

Although the potential shifts of the lanthanide ions can be measured by examining changes in the kinetic parameters of the membrane channels, an alternative approach is possible. The kinetics of the channel can be observed when the external solution has no trivalent ion and an easily measurable kinetic parameter such as the time to maximal current or the time to half-maximal current is recorded. A solution containing the ion then replaces the external solution. As the ions hyperpolarize the membrane, the observed times will increase significantly. These hyperpolarizing potentials are then compensated by applying an additional depolarizing potential to the membrane until the kinetic parameter again matches that of the original, lanthanide-ion-free solution. The additional depolarizing potential is then a direct measure of the magnitude of the hyperpolarizing shift induced by the lanthanide ions.

Such potential compensation experiments can be performed with the trivalent Er^{3+} ion on squid giant axon (Starzak and Starzak 1978). Hyperpolarizing shifts of 120 mV were observed 0.1 mM concentrations of this ion. The compensating depolarizing potential was applied and the kinetics were matched so that the maximum current appearance time corresponded to that of the original solution. Under such circumstances, both current records might be expected to match exactly. However, this was not the case. The observed maximal currents at the compensated, final potential were significantly larger than the corresponding currents for the original solution, even though both had the same kinetics. These data suggest that the gating process and the channels are functionally uncoupled. The maximal current is proportional to the ratio

$$\alpha/(\alpha + \beta) \tag{12.32}$$

However, since the kinetics in each case are adjusted for equality, this ratio must also be the same in each case and g_K is the same in each case. The additional current increase is produced entirely by the larger net potential difference necessary to restore the channel kinetics.

These experiments show that the gating process is distinct from the flow of ions in the channel. The lanthanide ions slow the gating kinetics but have no effect on the flow of ions through the open channel.

7. GATING CURRENTS

Because the gating molecules in the Hodgkin–Huxley model respond kinetically as a function of the transmembrane potential, they are very likely to be charged. When the membrane potential is changed, these molecules can then move or rotate in response to the resultant electric field. However, this suggests a motion of charge, which should give rise to a current. Hodgkin and Huxley originally proposed, for example, that such a gating molecule might have six charges that responded when the field was changed.

If there is a motion of charge, these resultant gating currents should be observable. Because they are expected to take place rather rapidly, it becomes necessary to separate them from the initial capacitative transient produced when the potential is changed at the beginning of the step potential clamp experiment. Because the capacitive response is linear, an exponential decay will be observed for either positive or negative potential steps: that is, if a positive potential step is applied, an exponential decay of outward current would be observed; if the potential step was negative, an exponential decay of inward current would be observed. If positive and negative potential steps were equal, the resultant current decays would be exact mirror images of each other. If they were subtracted, a net current of zero would be observed for the linear capacitative circuit.

To observe gating currents, an axon is voltage clamped with a sequence of pulses, such as potential steps of equal magnitude and opposite polarity. These pulse combinations are then "summed" by directing the resultant currents to a signal averager. The linear capacitative components will cancel, so that any residual currents can be attributed to other sources. Of course, for such experiments, the normal ionic currents must be eliminated because they are definitely nonlinear; currents are observed for depolarizing potentials, while hyperpolarizing potentials produce minimal currents. These ionic currents can be eliminated by introducing blocking agents that eliminate the sodium and potassium currents. The leakage component is linear, so it can be compensated.

When experiments such as these are performed on squid giant axon, residual outward currents are observed. These are shown in Fig. 12.9 with the axon's sodium current for comparison. The outward "gating" current rises rapidly and falls as the turn-on of the sodium current begins. Interestingly, the gating currents extend well beyond the range of the capacitative transient. Because the sodium currents can be quite large, it is crucial to eliminate these currents so that the gating currents are not obscured. This is accentuated by noting the difference in magnitude of the two currents. The sodium currents lie in the mA/cm^2 range, while the gating currents lie in the $\mu A/cm^2$ range.

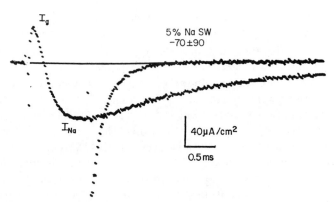

Fig. 12.9 Gating current (I_g) and sodium current (I_{Na}) recorded in 5% NaSW/290 CsF. Two traces have been superimposed. In one, the pulse (90 mV amplitude) was interrupted just after the peak of I_{Na}, and in the other it continued beyond the end of the trace. Each trace is the sum of current from five positive and five negative steps, 2°C. [Reproduced with permission from Benzanilla, 1974.]

8. THE COLE–MOORE SHIFT

Although the Hodgkin–Huxley equations predict the behavior of the bulk of the observations in squid giant axon, there are some experiments that are inconsistent with these equations. By focusing on these anomalous experiments, it is sometimes possible to gain new insights into the nature of the physical phenomenon being described.

In 1960, Cole and Moore (1960) performed a series of experiments in which the axon was preconditioned before it was voltage clamped to its final depolarizing potential. The axon was voltage clamped to a hyperpolarizing potential, such as − 100 mV, for a period of time, and the clamping potential was raised to a depolarizing potential so that ionic currents could flow. Since TTX was not available to block the sodium channels, Cole and Moore selected a depolarizing potential equal to the equilibrium potential of sodium so that

$$V - V_{Na} = 0 \qquad (12.33)$$

and no current could flow through the sodium channels. The residual sigmoidal current was the potassium channel current. When such experiments were performed for a series of preconditioning hyperpolarizing potentials, the resultant current curves were observed to shift in a parallel fashion, as shown in Fig. 12.10. As the initial voltage was made more hyperpolarizing,

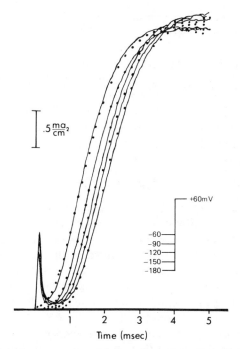

Fig. 12.10 Potassium current records in voltage-clamped squid axon. Each record is clamped to the same final potential, but increasingly negative preconditioning voltage pulses produce a parallel shift to longer times.

the resultant current records produced a longer induction period. After this induction period, the rapid rise of current paralleled the current record for the experiment in which the axon was not hyperpolarized.

At first glance, the explanation for this phenomenon would seem to be obvious. When the membrane is hyperpolarized, the fraction of gating molecules with "closed" configurations increases and the first-order process that dictates the evolution to the final steady state must evolve through a larger number of intermediate configurations.

When these data are analyzed with the Hodgkin–Huxley equations, the current shifts are found to be much too large to be fit by a simple adjustment of the initial fraction of open states. In fact, if the initial condition is reduced to $n = 0$ (no open gating molecules), most of the shifts cannot be fit using the Hodgkin–Huxley equations.

A number of theoretical models have been proposed to explain the magnitude of the Cole–Moore shift. One of the first was proposed by Hill and Chen (1971) and postulated interactions between the four units that controlled currents in one potassium channel. The qualitative picture is quite simple. If the molecules can interact, then certain configurations might be stabilized and the system would require a longer time to evolve to its final state.

The major question is then how to introduce the interactions into the kinetic scheme. The Hodgkin–Huxley equations can be replaced by a mathematically equivalent series of sequential first-order kinetic equations. Only the Hodgkin–Huxley first-order rate constants α_n and β_n are allowed, but different kinetic steps may have rate constants that differ from these by an integer factor. Thus, the sequence that reproduces the Hodgkin–Huxley equations can be written in the form

$$0 \underset{1\beta}{\overset{4\alpha}{\rightleftharpoons}} 1 \underset{2\beta}{\overset{3\alpha}{\rightleftharpoons}} 2 \underset{3\beta}{\overset{2\alpha}{\rightleftharpoons}} 3 \underset{4\beta}{\overset{1\alpha}{\rightleftharpoons}} 4 \qquad (12.34)$$

The subscripts have been deleted from these rate constants for convenience. State 4 corresponds to all four gating molecules being in their open configuration, so it is the state that will permit a flow of ions.

The sequential kinetic scheme provides additional clarity on how the presence of interactions might slow the appearance of open channels. If any of the rate constants for intermediate states are reduced by the interactions, the overall evolution to the final state must also be slowed.

However, the joint conditions of an inductive delay followed by a current rise parallel to the reference current rise (superposition) are extremely difficult to satisfy with most of the interactive models. If the rate constants in the sequence are altered to produce adequate inductive delay, the resultant currents will not remain parallel. If the equations are designed for optimal parallel behavior, the inductive characteristics are lost.

Hill and Chen concluded that such models were inadequate to describe the Cole–Moore data, and they proposed an alternative. The rate constants for the sequence in Eq. (12.34) were not changed. However, an additional kinetic process that preceded this sequence was postulated. When the membrane was hyperpolarized, this state was populated and the larger fraction of the gating molecules might now be found in this state. When the membrane was depolarized, the gating molecules evolved from this initial state to State 0, and from there they evolved through the normal Hodgkin–Huxley sequence of states. Because the new state was nonconducting, it functioned as a delaying state to produce the proper induction period. Because the system evolved into the Hodgkin–Huxley sequence after this time, it gave current curves parallel to the Hodgkin–Huxley current curves.

Because this new kinetic state cannot be observed directly, this hypothesis cannot be tested. Two possible explanations have been suggested. In the first case, phospholipid molecules may be absorbed or desorbed on the surface of the channel gating molecule. In the second case, the hyperpolarizing field pulls the charged gating molecule out of position. The delay is then produced by the time required for this gating molecule to return to its original position under the action of some restoring force on the molecule. Both models are plausible.

For the models discussed thus far, the kinetic steps are always first-order. An alternative approach leads to kinetics that are not first-order when there are interactions between the gating molecules for a channel. The differential equation for the fraction of open states,

$$dn/dt = \alpha(1 - n) - \beta n \qquad (12.35)$$

can be interpreted in two different ways. In the first case, the equation describes the kinetics of an ensemble of channels as a function of time. If this system were examined at some time t, then $n(t)$ would give the average number of all the gating molecules in the open configuration. If an accurate average were required, a large number of channels would be required or the system could be examined a large number of times. The "average" $n(t)$ for each experiment would then be averaged to get the best average value.

The alternative approach assumes that Eq. (12.35) describes the average kinetics of a single channel. In other words, the average is built into the equation initially. Each channel in an ensemble might be behaving slightly differently, but the channel describing the average behavior of all these channels is introduced and used to study the kinetic evolution of the system.

If the average channel kinetic approach is used, interactions between molecules can be introduced in an entirely different way. The kinetic equation for a single gating molecule is modified to include a dependence on the average probability that at least one neighboring molecule is already in its open configuration. The set of four closed gating molecules is stabilized by interactions between the molecules. If one of these molecules attains an open configuration, the remainder will be stabilized and their kinetics will reflect the presence of their altered neighbor.

In order to use this approach with the average channel formalism, it is necessary to determine the probability that at least one neighbor is in the open configuration when the fraction of open gating molecules is n. If $n \neq 0$, then one of the neighbors is expected to be open. However, in this average channel we are dealing with probabilities for the open configuration. In a nonprobabilistic system, when $n = 0.25$, then one molecule in each channel will be open and no interaction is possible. In this probabilistic case, however, at $n = 0.1$, this fraction will still exert an influence on the kinetics of the average channel.

The probability that at least one neighboring molecule is open can be determined in the following way. If there are z nearest neighbors, the probability that all z are closed is just

$$(1 - n)^z \qquad (12.36)$$

because each molecule is assumed to be in its configuration independently of the others. The probability that at least one is in the open configuration is then just

$$1 - (1 - n)^z \qquad (12.37)$$

The Hodgkin–Huxley differential equation for K channels is then

$$dn/dt = [\alpha(1 - n) - \beta n][1 - (1 - n)^z] \qquad (12.38)$$

With this equation, the fraction of open channels depends on both the normal first-order process in which each molecule changes independently of the others and the probability that at least one neighbor is already open. If the number of nearest neighbors is large, for example three, the second term in Eq. (12.38) approaches 1 very quickly and the equation reduces to the Hodgkin–Huxley equation. The nearest neighbor probability is important only in the hyperpolarizing regime.

9. ASCENDING POTENTIAL RAMPS

In most of the studies that we have discussed here, the voltage clamp involved a potential step: that is, the potential was changed to some new value and maintained there by the voltage clamp throughout the experiment. However, some interesting results are found if the membrane is clamped with some time-dependent potential functions.

An ascending potential ramp can be used as the clamping function. When a potential is applied with a small ramp slope such as 0.5 mV/ms, the resultant current record shows a delay period followed by a rapid rise in current. Because the slope of the potential function is constant, there is a direct proportionality between the time and the potential applied to the membrane. If the slope is S, then

$$V(t) = St \qquad (12.39)$$

Because of this, the current-versus-time curve that results from the experiments can be transposed directly into a current-versus-voltage curve by a simple change of units on the abscissa. The resultant current–voltage (I–V) plot is shown in Fig. 12.11. It is essentially equivalent to the curve that could be generated by plotting the steady-state current from a series of step potential clamp experiments against these clamping potentials. When the ramp slope is small, the ascending potential ramp reproduces the steady-state potassium channel I–V curve.

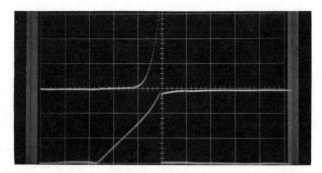

Fig. 12.11 Voltage clamp of squid axon with ascending potential ramps. Increasing ramp slope produces an enhanced induction period and a steeper current rise.

As the ramp slope is increased, the $I-V$ curve shows an increasing inductive period before the current begins to rise. The current then rises more steeply, so that all curves reach their same final steady state, that is, the current for the maximum potential of the ramp.

There is a logical reason for the delayed rise. If all the channels were initially in their closed configuration and a potential ramp was applied, the first-order equation for the process would become complicated because the rate constants depend on the potential. As the ramp potential changed, so would the rate constants. The first-order equation would have the form

$$n(t) = n_{ss}\left(1 - \exp\left\{- \int [\alpha(V) + \beta(V)]\, dt\right\}\right) \qquad (12.40)$$

For an ascending potential ramp, the early voltages are small and the rate constants in the integral are small, so that the exponential term decays very slowly and the current remains quite small. As the potential increases, the exponential decays very rapidly and the current rises rapidly toward its final steady-state value.

PROBLEMS

1. What are the advantages of voltage clamp experiments on nerves relative to simple stimulus (current clamp) experiments?

2. Describe the expected effects due to the following experiments on an excitable membrane: (a) a decrease in the external sodium concentration followed by a depolarizing voltage clamp potential; (b) a decrease in the external potassium concentration followed by a voltage clamp pulse; (c) a stimulating current less than the excitation threshold; and (d) A hyperpolarizing potential applied before a depolarizing potential.

326 12. *Membrane Excitability*

3. An alternate method of solving the Hodgkin–Huxley equations for n is

$$dn(t)/dt = \alpha_n(1 - n) - \beta_n n = \alpha_n = \alpha_n - (\alpha_n + \beta_n)n = y$$

Using this method, (a) determine dn in terms of dy; (b) generate a new differential equation in terms of y and solve it; and (c) determine $n(t)$.

4. Assume a nerve is selectively permeable only to Na^+ ions as a function of time, and that K^+ ion permeability does not change. How would the action potential and voltage clamp currents change for this system?

References

Basano, L., and Ottonello, P. (1981). *Am. J. Phys.* **49**, 672.

Benzanilla, F. (1974). *J. Gen. Physiol.* **63**, 533.

Clark. H. T. (ed.) (1954). "Based upon a Symposium on the Role of Proteins in Ion Transport across Membranes, Sponsored by the NSF," D. Nachmansohn, Assoc. Ed. Academic Press, New York.

Cole, K. S. (1932). *J. Cell. Comp. Physiol.* **1**, 1.

Cole, K. S. (1968). *From* "Membranes, Ions and Impulses." Univ. of California Press, Berkeley.

Cole, K. S., and Moore, J. (1980). *Biophys. J.* **1**, 1.

Danielli, J. (1933). *J. Cell. Comp. Physiol.* **2**, 75.

Frankenhaeuser, B., and Hodgkin, A. (1957). *J. Physiol.* (*London*) **137**, 218.

Glansdorff, P., and Prigogine, I. (1971). "Thermodynamic Theory of Structure, Stability and Fluctuations." Wiley-Interscience, London, New York.

Gorter, E., and Grendel, F. (1925). *J. Exp. Med.* **41**, 439.

Hill, T., and Chen, Y. (1971). *Proc. Natl. Acad. Sci. USA* **68**, 2488.

Hodgkin, A., and Huxley, A. (1952). *J. Physiol.* (*London*) **117**, 500.

Hodgkin, A., and Keynes, R. (1955). *J. Physiol.* (*London*) **128**, 61.

Keynes, R. (1955). *J. Physiol.* **128**, 61.

Lucy, J. A. (1968). *In* "Biological Membranes," (D. Chapman ed.). Academic Press, New York.

Moore, W. (1955). "Physical Chemistry, 2nd Ed." Prentice-Hall, Englewood Cliffs, New Jersey.

Moore, W. (1972). "Physical Chemistry, 4th Ed." Prentice-Hall, Englewood Cliffs, New Jersey.

Ovchinnikov, Yu. A. (1974). *From* "Membrane Active Complexones." Elsevier, Amsterdam.

Overton, E. (1899). *Viertjahrsschr. Naturforch. Ges. Zuerich* **44**, 88.

Parlin, R., and Eyring, H. (1954). *In* "Ion Transport in Membranes," (H. T. Clark ed.). Wiley, New York.

Plowe, J. (1931). *Protoplasma* **12**, 196.

Rosenberg, R., and Finkelstein, A. (1978). *J. Gen. Physiol.* **72**, 327.

Schmidt, W. J. (1936). *Z. Zellforsch. Mikrosk. Anat.* **23**, 657.

Starzak, M., and Starzak, R. (1978). *Biophys. J.* **24**, 555.

Stein, W. D. (1967). "The Movement of Molecules across Cell Membranes." Academic Press, New York.

Temple, P. (1975). *Am. J. Phys.* **43**, 801.

Index

A

Action potential, 83, 302
Activity, 70
  ions in membrane, 84–86, 225
Activity coefficient, 70
Admittance, 150
Advancement, 247
Affinity, 247, 263
Agar, 108
Agitation, thermal, 198
  dipoles, 57–58
α, Hodgkin–Huxley equations, 313, 315
Amphipathic, 4
Amplifier
  difference, 122
  exponential, 127
  inverting, 111, 133
  noninverting, 121
Ammeter, 119
Analog to digital converter, 136
AND, 133, 135
Area, per molecule, 9
Arrangements
  molecular, 6, 34, 37, 42
  ionic, 178
Avogadro's number, 9

B

Balance, surface, 7
Belousov–Zhabotinsky reaction, 260
β, Hodgkin–Huxley equations, 313, 315
Binding
  average, 40–41
  equilibrium, 42, 192

Birefringence, 6
Boltzmann constant, 25
Boltzmann factor, 33–42
Boltzmann statistics, 33–42, 152, 180
Boundary conditions, 165–166, 170–171, 182, 214, 220
Bragg relation, 15
Bridge, 152

C

Capacitance, 60–61
  cell suspensions, 164–172
  differential, 191, 197
  integral, 191
  membrane, 141
  stray, 141
Capillary, 195, 198, 201
Carrier, 287
Centrifuge, 13, 88
Ceric–cerous ions, 260
Channels, ionic, 198, 230, 292, 294, 310, 314
Charge
  cgs units, 45
  image, 62–64
  MKS, 44
  surface, 31, 177, 195
Charge flux, 219, 224, 239, 250, 258
Charge transfer
  electrode, 92
  concentration cells, 104, 106
Cholesterol, 4–5
Choline, 304
Cole–Cole plot, 153–161
Cole–Moore shift, 320
Complex conjugate, 149

329

Concentration, surface excess, 32–33, 69
Concentration cells, 103–106
Conductance, 100
Conductivity, 101, 222
    molar, 101
    equivalent, 101, 103, 171
Constant field, 220, 227
Converter, current to voltage, 120, 133
Coulomb law, 44, 48
Coupling, steady state, 262–265
Curie symmetry principle, 259–261
Current–voltage curves
    instantaneous, 307
    steady state, 309
Cylinder, 46, 185

D

Debye unit, 50
Debye length, 181, 183
Debye relaxation time, 161–163
Density
    charge, 46, 179, 181, 183
    molecular, 9
Detailed balance, 284
Dielectric constant, 17
    real, imaginary components, 156–161
    relative 17, 61
Differentiator, operational amplifier, 124–127
Diffusion
    discrete state, 230
    membrane, 213
    rotational, 162
    time dependent, 213–217
Diffusion coefficient, 209, 219
    discrete state, 234
Digital to analog converter, 139
Diode, 127
Dipole
    electric, 48–60
    induced, 58–61
    rotation, 52–53
Dipole moment
    average, 36, 39, 57
    relaxation, 158
Dipole–dipole interaction, 53
Displacement, electric, 158
Dissipation, resistive, 142, 243
Dissipation function, 243–247, 261
Dissociation, ionic, 74

Divergence, 47, 213
Divergence theorem, 47
Donnan equilibrium, 75–79, 253
Double layer, 176, 188–193, 200, 317

E

Electrode, 91
    Ag, 92
    Ag–AgCl, 93
    calomel, 95–96
    gas, 92
    Hg, 95, 196
    nonpolarizable, 94
    polarizable, 94
    Pt, 91
    second kind, 93
Electrolyte, z:z, 186
Electroneutrality, 77, 175, 181
Electroosmosis, 198–201, 250
Electrophoresis, 198–201
Energy,
    activation, 273–275
    average dipole, 35, 38
    centrifuge, 89
    dipole–dipole, 55
    dipole in field, 53
    induced dipoles, 60
    internal, 22
    spherical cell in field, 168
    surface area, 11, 28
Energy barriers, 231–236, 274, 275–277
Enthalpy, 28
Entropy, 24–27
    activation, 278
Equation of continuity, 213
    entropy, 245
Equilibrium constant, 75, 192, 272
    activation, 275–277
Equivalent circuit, 142–143, 145–146
Euler theorem, 31
Expansion, 23
Eyring, 231, 275

F

Faraday, 80
Feedback, 116, 121, 133
FET, 116

Fick's first law, 209, 240, 255
Fick's second law, 211–213
Field
　electric, 46, 165, 185, 239
　electric dipole, 51
　electric vector, 17–18, 239
Filter, RC, 140
Flux
　charge, 219, 222, 239–240
　generalized, 240–242
　heat, 208
　particle, 209
　scalar, 246
　solute, 255
　solvent, 255, 258
Flux ratio, 227–229
Force
　centrifuge, 90
　driving, 200
　duNoüy ring, 12
　frictional, 199, 202
　generalized, 240–242
　surface, 7
Force–flux relations, 241, 249, 253
Fourier law, 208, 242
Fourier series, 216
Free energy, Gibbs, 29, 34
　activation, 275
　electrochemical, 29
Friction, 199, 202
Frictional coefficient, 199
Fugacity, 67
Function generator, 135

G

Gain, 116
Gating molecule, 310
　currents, 319
　interactions, 321–324
Gauss' law, 47, 185
Gibbs adsorption isotherm, 31–33
Gibbs–Duhem equation, 32
Gibbs–Helmholtz equation, 274
Goldman equation, 220–223, 234
Goldman–Hodgkin–Katz equation, 224–226
Gorter–Grendel experiment, 10–11
Gouy–Chapman theory, 178–184
Gradient
　concentration, 255

potassium, 84
sodium, 83
thermal, 244, 262
Grahame equation, 184–188
Gramicidin, 258–259
Graph theory, 281–282
Gravitation, 12, 88

H

h, Hodgkin–Huxley equation, 315
Heat, 21
　reversible, 24
Helmholtz double layer, 177, 183, 200
Helmholtz plane, 188–192
Henderson equation, 109–112
Hill-Chen model, 321–322
Hodgkin–Huxley K^+ equation, 310–314
Hodgkin–Huxley Na^+ equation, 314–317

I

Ideal gas, 30
Impedance
　components, 149, 153–156
　input, 114–115
　output, 115–116
Independence, 304
Index of refraction, 16
Information, 27
Integrating factor, 143, 145, 158, 311
Integrator, operational amplifier, 123–126
Interactions, 53, 322–323
Interface, electrode, 91–93
Ions, polyvalent, 78, 82, 182, 186, 261, 317
Ion repulsion, 296
Ion transport, channels, 230, 292, 294
Ionic strength, 181

J

Junction, summing, 119
Junction potential, 97

K

Kinetics
　cyclic, 284–287

first order, 267
first order, reversible, 268
Michaelis–Menton, 278–280
nonsteady state, 298–300
passive transport, 287–291
reciprocal plot, 280
second order, 270
transport, channel, 292–298

L

Langevin equation, 36–39
Lanthanides, 59, 318
Laplace equation, 164
Laplace transform, 299
Leakage, 316
Lettvin axon, 131, 132
Light, polarized, 17
Lines of force, 167
Lipid, 3–4
Lippman equation, 30, 195
Lock-in amplifier, 139–140
Logic, 132–134
Long pore, 229

M

m, Hodgin–Huxley equations, 314
Mass, in centrifuge, 13, 90
Maxwell relations, 30, 66, 258
Micelles, 10
Microbeam, 13
Mole fraction, 67, 71
Monolayer, 7–9
Monostable multivibrator, 134
Multistate kinetics, 230, 281–300, 322

N

n, Hodgkin–Huxley equations, 311
NAND, 133
Nernst–Planck equation, 112, 218–220, 226
Network
    parallel, 144, 155
    RC, 141–146
    sinusoidal potential, 146–150
    series, 143

Noise, 126, 136, 137
    pseudo-random, 153
NOR, 133

O

Ohm's law, 208, 219, 238, 240
Open collector, 133
Operational amplifier, 116, 133
OR, 133
    exclusive, 133
Osmotic mobility, 251
Osmotic pressure, 70–74

P

Parabola, 198
Parlin–Eyring, 230–234, 275
Partition function, 35, 38
    grand canonical, 39–42
Permeability coefficient, 222
pH, 86–174
Phase delay
    electrical, 148
    light, 18
    lock-in, 140
    x-ray, 15
Phosphatidic acid, 3
Plowe, 1
Poise, 201
Poisson equation, 47, 179
Poisson–Boltzmann equation, 181, 185
Polarizability, 58
Potential
    cell suspension, 166
    chemical, 29, 209, 245
    control, 133, 303
    depolarizing, 313
    diffusion, 107–109
    dipole–charge, 48
    dipole–dipole, 53–56
    electrochemical, 79, 80, 86, 103, 225
    Galvani, 99
    hyperpolarizing, 313
    interfacial, 97–99
    reference, 91
    transmembrane, 81–83, 195, 223
    Volta, 98
    zero charge, 197

Pressure
  external, 23
  hydrostatic, 13, 253
  surface, 7
  transmembrane, 250
  vapor, 67
Probability, 25, 323
  Boltzmann, 34–35
  weighted, 25
Projection, 17
Protein, 6

R

Ramps, potential, 125, 324
Raoult's law, 67
Reaction, force–flux for, 246–249
Reciprocal relations, 252
Rectifier, 223
Reflection coefficient, 257
Regulator, voltage, 136
Resistance, 100
  measurement, 150–153
  membrane, 141
  parallel, 145, 150
  series, 133, 143, 155
Reversible, 23

S

Salt bridge, 108
Sample–hold, 137
Saxen relations, 257–259
Screening, 192–195
Sedimentation potential, 204
Semicircle, 154–156, 159–160
Separation of variables, 215
Series resistance compensation, 133
Servomotor, 135
Signal averaging, 136–139
Signal-to-noise ratio, 137
Steady state, 279
Stern model, 188
Steroid, 4–5
Stokes law, 163
Streaming current, 203, 251
Streaming potential, 203, 251, 259
Summer, 119–121
Surface excess, 33

T

Taylor series, 73
Temperature, external, 24
Tensiometer, 12
Tension, surface, 11–14, 195
  cells, 12–14
Test charge, 48–49, 174
Tetraethylammonium (TEA), 304
Tetrodotoxin (TTX), 304
Thermodynamics, first law, 22
Time constant, 124, 143, 150
Timer, 135
Torque, 52, 161
Transient, capacitative, 126, 144
Transition state theory (*see* Eyring)
Transport number, 102
Trigger, 125, 133

U

Ultracentrifugation, 88–90

V

Valinomycin, 288
Velocity
  angular, 89
  in electric field, 103
  shear, 199
Virtual ground, 119, 133
Viscosity, 163
Voltage clamp, 128–131, 302
  action potential, 131
Voltage follower, 122
Volume
  molecular, 9
Volume, partial molar, 72

W

Waage–Guldberg, 272, 274
Wegscheidner, 287
Work
  PV, 21
  reversible, 23

X

x–y recorder, 125
X-ray, 14–16

Z

Zeta potential, 200, 203, 251